AF633039

ENERGY MANAGEMENT
CHALLENGES FOR THE NEXT MILLENNIUM

By the Same Author

Bio-Energy for Rural Energisation

Bio-Energy Resources: Planning, Production and Utilization

Energy Management: Policy, Planning and Utilization

Energy Management: Challenges for the Next Millennium

Financing of Energy Sector in Development Countries

Food Security and Panchayati Raj

Strategy for Energy Conservation in India

Sustainable Energy Supply in Asia (in 2 vols.)

Women and Food Security: Role of Panchayats

About the Author

Mr. Pradeep Chaturvedi is an expert on energy management, renewable energy and rural energy. His expertise lies in project planning, implementation and management with emphasis on the South Asia Region. A number of pioneering projects and studies of national and international significance have been conducted under his guidance. His involvement with the FAO's Regional Wood Energy Development Programme, based at Bangkok, has given him exposure to biomass use for energy in the South Asia Region. He is currently working as a Consultant to UNIDO's International Centre for Application of Solar Energy.

Mr. Chaturvedi is a Member of the Indian Member Committee of the World Energy Council and is responsible for coordinating India's programme with WEC.

Mr. Chaturvedi is also Vice-Chairman of the UNESCO sponsored World Renewable Energy Network based in UK; President of the Bio Energy Society of India and Secretary General of the Indian Association for the Advancement of Science. He is also a member of the Rural Energy Study Committee of the World Energy Council. Mr. Chaturvedi is a graduate in Engineering. He is a Fellow of the Institute of Directors; Fellow of the Institution of Engineers (India) and Chairman, Mechanical Engineering Division Board.

He is also the recipient of FAO's World Food Day Medal, 1991 and selected for the 'Pioneer in Renewable Energy' Award of the World Renewable Energy Network, U.K. for 2000.

Mr. Chaturvedi has edited and authored 29 books related to Energy Management, Rural Energy, Energy Conservation and Sustainable Development.

ENERGY MANAGEMENT
CHALLENGES FOR THE NEXT MILLENNIUM

PRADEEP CHATURVEDI

CONCEPT PUBLISHING COMPANY, NEW DELHI - 110059

All rights reserved. No part of this work may be reproduced, stored in a retrieval system, or transmitted in any form or by any means, electronic, mechanical photocopying recording or otherwise, without the prior written permission of the copyright owner and the publishers.

ISBN No 81-7022-806-9

First Published 2000

© Author

Published and Printed by
Ashok Kumar Mittal
Concept Publishing Company
A/15-16, Commercial Block,
Mohan Garden, New Delhi-110059

Phones : 5648039, 5649024

Fax : 091-11-5648053

Dedicated to

P.J. ADAM

Chairman

World Energy Council Executive Assembly

for

his commitment to the sustainable supply and use of energy for the greatest benefit of all; encouragement to linkage of business interests to the reduction of poverty amongst 2 billion poor people without access to commercial energy; and vision to lead the World Energy Council into the Next Millennium.

कृष्ण चन्द्र पन्त
K.C. PANT

सत्यमेव जयते

उपाध्यक्ष
योजना आयोग
भारत
DEPUTY CHAIRMAN
PLANNING COMMISSION
INDIA

May 18, 1999

FOREWORD

The challenge before the planners is to ensure sustainable and cost effective energy supply, keeping in mind the synergies between energy, environment and development, so that the energy needs of the people are met.

This book by Shri Pradeep Chaturvedi is a good attempt to cover various aspects of the energy scenario at the micro, national and global levels. I am happy that the issues of demand side management through energy conservation and of augmenting environmentally friendly energy supply have been well integrated. It gives an over view of the energy sector, covering, interalia, policy reforms and future options for energy technology.

I am sure that this book would be useful for policy makers, researchers, implementing agencies as well as for the general reader who is interested in the energy sector. I wish Shri Chaturvedi all success in this venture.

(K.C. PANT)

योजना भवन, संसद मार्ग, नई दिल्ली – ११० ००१ दूरभाष (कार्यालय) : ३७१५३७६, ३७१६३५४ फैक्स : ३७२३५८७
Yojana Bhawan, Parliament Street, New Delhi - 110 001 Phones (Office) : 3715376, 3716354 Fax : 3723587
दूरभाष (निवास) : ३०१४४३१, ३०१२६१८ फैक्स : ३०१६४५३ Phones (Resi.):3014431, 3012618 Fax : 3016453

PREFACE

Energy is an essential ingredient of human life on our planet today. Just as human beings must have air, water, food, clothing and shelter, they must have energy for survival. There is no turning back the clock to prehistoric times when human beings were able to exist without energy. The conditions of those times would not support today's global population.

At the same time, we live on a planet where two billion people have no access to commercial energy. Those same people also live in poverty and, while there is not necessarily a cause and effect relationship, the solutions which will raise them up out of poverty will also bring them commercial energy. We need to understand what is required and work diligently to solve the twin problems of poverty and lack of energy.

India, the second most populous nation in the world, can serve as a good model for our solutions. Efforts within and outside of India are needed to improve conditions for the people of India. Without energy, poverty will continue, and without our efforts there will be no additional energy.

We must provide the beacon, which will guide the world's elected leaders on a path of increasing energy availability coupled with bringing people up out of poverty; a path marked by improved efficiency, the application of sound technologies, and care for our environment. As Chairman of the WEC Executive Assembly, I proposed a five-point programme to our 17th Congress in Houston in September, 1998.

First, we must recognize that economic growth in both developing and industrial countries will be essential to human development and, with that economic growth, will come demand for increasing, affordable energy resources.

Second, we must continue to pursue improved efficiency in both production of energy and use of energy.

Third, we must be keenly aware of the large amount of finance needed by the global energy industry. WEC studies indicate the finance is available, but we need to better understand and promote the policy options, which will attract the needed finance.

Fourth, we must protect our environment. We need to demonstrate that aggressive environmental protection costs less, not more; provided it can be done within a rational framework of governmental policy. Further, we must provide more light on that rational path to environmental protection, so our elected leaders can more easily follow the path.

Fifth, we must recognize that, even in the absence of ratification of the Kyoto agreement, some limit on atmospheric CO_2 concentration is likely to be established through a consensus of most governments, including those of developing countries. Hopefully, any agreement will include provisions for emissions trading, because those provisions tend to lead to optimum solutions.

As we address a world in which economies are growing and energy resources are expanding to meet increased demand; yet CO_2 emissions must be limited, we will be forced to confront the truth.

New renewable energy resources including solar, wind and bio-mass must be expanded as rapidly as possible. We need to better understand how much production we can get from these resources.

The old renewable energy resource, hydroelectric power must be expanded, particularly in developing countries, where there is tremendous potential. We will have to deal with the dichotomy of the environmental degradation, which may result from damming our rivers, balanced by the environmental enhancement of a resource, which produces no CO_2.

We must emphasize natural gas for electricity production, because it produces about one-half the CO_2 of a comparable coal plant.

Hydrogen is a fuel, which can be burned without CO_2 production. We must learn more about the contribution hydrogen technology can make to our quest.

Information technology has a role to play and we need to understand it better.

Nuclear energy holds tremendous promise as a non-carbon emitting resource. We must continue to run the plants we have and must build more. The issues of cost, waste isolation, non-proliferation, and public perception of nuclear safety must be resolved. A world in which CO_2 production must be limited provides the strong incentive we need to resolve all of these issues.

I hope these ideas, considered in the context of this book by Pradeep Chaturvedi, will provide a framework for progress in making more energy available to the people of India. Certainly, the book itself provides a great deal of important information to get us started on the path to making energy accessible to all.

Date: April 30, 1999

P.J. Adam

Chairman of the Board, Black and Veatch Holding Company, and
Chairman of the World Energy Council Executive Assembly

CONTENTS

INTRODUCTION

India meets its energy needs at a low level of consumption, but, is not as comfortable as it would like to be. The global recession and its adverse impact on India's economy has reduced the energy demand. National level consultations have identified new directions for developing the energy sector to meet all the needs. Measures have to be evolved to counter Major Challenges of the energy sector for smooth transition to the Next Millennium. These challenges can be identified as follows:

Major Challenges of the Next Millennium

Providing Access to Energy Source to the Poor

The primary concern of the government is the welfare of the poor and needy people. Every effort is necessary to provide better quality of life to them. Meeting their energy needs is an important element of the policy. Energy consumption in the household sector has been ever increasing with the improvement of incomes of the people. In spite of this shift, over 350 million people under the poverty conditions do not have access to commercial sources of energy. Traditional knowledge and instinct has made them use traditional biomass to meet their subsistence energy needs. Government has introduced technological innovation supporting efficiency improvement. Emphasis has also been placed on other renewable energy sources to meet the energy needs of the poor rural people. New technologies are converting solid biomass fuel into gaseous fuel. Further advancement through the private sector ventures can certainly help in meeting future needs of the rural people for domestic as well as income generation needs. Majority of the poor population relies on agriculture for their earning. Introduction of solar energy based devices like solar greenhouse and solar chicken brooder have proved to be a viable income generating devices for the rural areas.

Accelerate the Use of Traditional Biomass and New Renewables

Use of traditional woodfuels will continue to the year 2020 in the developing countries, mainly in the South Asia, and India is no exception. Dedicated energy plantations will ensure sustainable woodfuel supply. Recent advancements in improving efficiency of combustion of biomass and their introduction in the development world have opened avenues for environment friendly technologies. At the same time, a new thrust on conversion of biomass into electricity, gas and liquids rather than its use as a solid biomass improved its acceptability as the final form of energy. With limited oil resources in India, and the rapidly increasing rural population, it will be necessary to develop new technologies and practices for higher use of traditional biomass in India. At the same time, new renewables have to be developed through intensive R&D.

Energy Sector Reforms

India's commitment to energy sector reforms have taken deeper roots. Government's role as the main funding source through budgetary allocations and through the government controlled financing agencies makes policymakers cautious about opening out to the private and foreign investors at a fast pace; and therefore, the public sector dominance continues . The public sector units have realised that they need to improve their efficiency of operation and become commercially viable and competitive. This culture can be introduced only gradually. They need support from the consulting organisations that can serve as bridge between them and other major players globally. A number of innovative human capital development programmes have been initiated in the public sector organisations. New strategies are now applied by them. A shift in the approach of the public sector organisations is apparently visible. The private sector on the other hand is cautious. Though they are interested in entering the market with assured and reasonable returns they prefer to let the public sector units show the path. Energy sector reforms have to support Mega Power Projects, as well as decentralised small systems.

Energy Conservation and Efficiency Improvement

Efficiency improvement, and thereby the plant load factor improvement, of the thermal power plants have contributed significantly in meeting energy needs from energy supplies from the same source. However, process is approaching the threshold limit. Recent observations (during the year 1997-98) have indicated a reversal of improvement in power plant load factor. That just indicates reaching a plateau and no further possibility of an increased output of more useful power from the same installed capacity. Energy conservation efforts now appreciated in industry, agriculture and domestic sectors need be intensified.

Energy and Environment Interaction for Controlling GHG Emissions

Kyoto Protocol (1996) and its follow up has set in motion serious discussions at the international level on containing (GHG emissions, even though the final ratification will be over only in December, 2000. Clean Development Mechanism is to be delivered by January 1, 2000 to put in place a globally acceptable flexible funding mechanism to fight GHG emissions. Different models are already emerging on "emissions trading" and "clean development mechanism". India is not faced with the threat of high GHG emissions, but it is necessary that steps to ensure clean environment continues. Emissions from automobiles, thermal power plants and industries cause intense localised pollution from energy use, and need to be appropriately handled. India has to carefully select from the available and emerging technology options.

Information Technology and Quality Energy

Introduction of efficient manufacturing and service sector systems and latest information technology based instruments make it necessary that the energy delivery, mainly in the power sector, is maintained at the optimum level. Voltage fluctuations have been causing permanent damage and lowering the reliability. A conscious effort is encouraged to provide quality energy services so that the best efficiencies are built into the industry, agriculture and service sectors. The reliability of energy systems and optimised operation of the communication networks can also lead to reduction in travel and hence the stress on the transport system.

Regional Cooperation

Regional cooperation in sharing of resources and energy services will be mutually beneficial to India and other neighbouring countries. The challenge here is to develop a mechanism that will have a long term workable acceptance. India has treaty with Nepal and Bhutan to buy surplus hydro-electricity. A number of potential sites exist in Nepal and

Bhutan that can be further developed. But, the developer needs guarantee for repayment of his investment that can be collectively given by the beneficiary countries. Transmission of such power will be facilitated by the national power grid being set up by the Power Grid Limited. It has plans to evacuate power from Nepal, Bhutan, Bangladesh and Pakistan. Every neighbouring country has to be convinced of the financial benefits and long term tenability of such arrangement.

The Book

The Book deals with the above mentioned challenges, mainly in the Indian context. Various national and global level consultations provided me an opportunity to present my opinion and synthesise on the basis of the feedback from other participants. The Book, accordingly, is based on such presentations after certain editing. The first Chapter on Energy Scenario in India presents the status as in March 1999. Other Chapters include suggestions to tackle the situation. Annexures have been added to include latest information for the reader.

Acknowledgement

The Foreword for the Book from Hon'ble Mr. K.C. Pant, the Deputy Chairman, Planning Commission of India reflects his perception of the energy sector based on his lifetime involvement at the highest level of governance and provides guidance for transition to the Next Millennium

The Preface for the Book from Mr. P.J Adam, Chairman, World Energy Council Executive Assembly, who has the responsibility to lead the Global Energy Community into the Next Millennium, reflects his deep concern for providing energy source to 2 billion poor in the developing countries.

Blessings from Mr. K.C. Pant and Mr. P.J. Adam add to the value of the Book and enrich its contents.

Dedication

The concern of Mr. P.J. Adam for fulfilling the energy needs of the 2 billion poor people, without access to commercial energy, has encouraged me to dedicate this Book to him, and look forward to his pursuing this cause vigorously.

New Delhi, June 15, 1999 PRADEEP CHATURVEDI

ACKNOWLEDGEMENTS

I wish to gratefully acknowledge the contribution of Foreword by Hon'ble Mr. K.C. Pant, Deputy Chairman, Planning Commission of India.

My grateful thanks to Mr.P.J.Adam, Chairman, World Energy Council Executive Assembly for his contribution of Preface and permission to dedicate the Book.

Permission from the World Energy Council and the FAO's Regional World Energy Development Programme in Asia has helped in including authentic information with due recognition of source.

I am also grateful to Mr. J.N. Maggo, Joint Adviser, Planning Commission; and Mr. L.A. Venkatesan, Central Electricity Authority for providing the help and inputs.

Mr. Ashok Mittal, Managing Director, Concept Publishing Company deserves all appreciation for designing the layout and publishing the Book.

PRADEEP CHATURVEDI

New Delhi
June 15, 1999

All rights reserved. No part of this work may be reproduced, stored in a retrieval system or transmitted in any form or by any means, electronic, mechanical, photocopying, recording or otherwise, without the prior written permission of the copyright owner and the publishers.

ISBN No 81-7022-806-6

First Published 2000

© Author

Published and Printed by
Ashok Kumar Mittal
Concept Publishing Company
A/15-16, Commercial Block,
Mohan Garden, New Delhi-110059

Phone : 5648039, [illegible]

Fax : 091-11-[illegible]

1

ENERGY SCENARIO IN INDIA

Introduction

Energy is the central element in the process of development. Realizing this aspect, energy development has attracted large public funding in the planned growth of the economy. Population growth rate had reached a threatening dimension and is decelerating only at a gradual pace. Efforts to promote fossil fuels and new and renewable sources of energy in an efficient fashion have ensured a minimum supply of energy in the country. The decade of 90s has witnessed gradual entry of the private sector companies and restructure of the public sector companies. The transition is accompanied by teething troubles and adjustment to changing circumstances. As the private sector will gain confidence, a number of smaller capacity decentralized energy systems will be set up.

Economic Situation in India

India covers an area of 328 million hectares, with an estimated population of over 965 million in 1997. Over 25 per cent of the population lives in urban areas.

The country witnessed a fast pace of economic reforms in the early part of this decade, followed by a period of stabilisation. International turbulence and unfavourable international economic environment also had its impact on economy during 1998-99. Extension of South East Asia crisis to countries in other continents resulted in a slow down of global growth to less than 2 per cent in 1998. World trade growth also decelerated sharply, commodity prices fell and deflation affected much of the global economy.

India's economy grew at an average rate of 6.8 per cent during the Eighth Five Year Plan period (1992-1997). During the terminal year of the Plan, GDP growth touched a phenomenal 7.5 per cent. There has been a certain deceleration to 5 per cent in 1997-98. The economy recovered to an estimated growth of 5.8 per cent in 1998-99 (estimated for the period April - December 1998). Agriculture and allied sectors have registered a growth rate of 5.7 per cent; electricity, gas and water supply 6.3 per cent; and mining 0.1 per cent. Trade, hotels and transport and communication accelerated to a growth rate of 6.8 per cent. Inflation rate rose sharply during 1998-99 because of an exceptional spurt in prices of a handful of agricultural commodities touching a peak of 8.8 per cent in late September, 1998, and then dropping to 5 per cent in early January, 1999.

The total gross domestic savings declined to 23.1 per cent of GDP in 1997-98 from 24.4 per cent of GDP in 1996-97. The real gross domestic formation was 25.6 per cent of GDP in 1997-98. Real gross fixed capital formation remained unchanged at 23.6 per cent of GDP.

The developments in the last few years have sent a strong signal that the economic reforms have come to stay. The 'pace' and 'pattern' may vary with political changes, but the broad thrust of liberalisation will not be reversed. The unleashing of the economy is generating its own momentum. Economics cannot be isolated from political decision making. The 1998 elections have shown that India has made a smooth transition from single party dominance to coalition governance with a tilt towards greater cooperative federalism and this could serve as a valuable corrective to earlier more centralising trends and could be a stabilising factor for the market economy in India.

High growth rate during 1992-97 brought to surface the infrastructural bottlenecks, especially in the power sector.

Distribution of Energy Resources

Primary Energy Resources and Regional Distribution

Coal and hydroelectric resources are the mainstay for India's primary resource support for secondary energy purposes. Oil and natural gas resources of the country are limited. Nuclear resources, on the other hand, are modest. The coal resources are of poor quality having high ash content. Besides, the distribution of the primary resources in the various regions is rather skewed. While the eastern region accounts for 70 per cent of the total coal resources, the western region accounts for over 70 per cent of the hydrocarbon reserves, and over 70 per cent of the total hydro potential of the country is concentrated in the northern and north-eastern region. The southern region which has only 6 per cent of the coal reserves and 10 per cent of the hydroelectric potential, possesses most of the lignite deposits of the country.

The regional distribution of the primary commercial energy resources are given in Table 1.1

TABLE 1.1 : Primary Commercial Energy Resources (March 1997)

Regional	Coal	Lignite	Crude Oil	Natural Gas	Hydro Power	
					Potential*	Energy
	(BMT)	(BMT)	(MMT)	(BCM)	(MW)	(TWH)
Northern	1.1	1.1	—	4	30135	225.00
Western	48.2	0.5	584	497	5679	31.40
Southern	13.1	25.9	—	—	10763	61.80
Eastern	141.4	—	—	—	—	5590
North-Eastern	0.9	—	148	159	31857	239.30
All India	**204.7**	**27.5**	**732660**		**84044**	**600.00**

Source: Planning Commission, Government of India (1998); and ONGC(1999).

BMT = Billion Metric Tonnes. MMT = Million Metric Tonnes.

BCM = Billion Cubic Metre.

* Potential at 60 per cent load factor.

Note: Gas reserves are now identified at 717.3 BCM (at the end of 1998) distributed under control of ONGC - 525 BCM; OIL - 94.5 BCM & Joint Ventures/Private companies - 97.5 BCM.

Non-conventional Energy Sources

The country is endowed with a large potential of non-conventional energy resources are given in Table 1.2.

Trends in Energy Production, Supply and Consumption

Indigenous Primary Energy Production

India's energy requirements are met from traditional as well as modern forms of energy. The share of traditional forms of fuels is diminishing, and in 1996-97 these contributed to only 34 per cent of the annual energy consumption. The usage is largely restricted to domestic fuels in semi-urban and rural areas at very low efficiencies.

Total power generation capacity installed on March 31, 1998 was 89,090 MW. (capacity addition of 6,324 MW is estimated for the period April 1, 1998 to March 31, 1999). These plants generated a total of

TABLE 1.2: Renewable Energy Potential In India (in March 1999)

Source/Technology	Potential	Availability	Potential Exploited
Biogas Plants	12	Million	2.713
Million			
Biomass-based Power	17,000	MW	114 MW
Efficient Woodstoves	120	Million	2.8.5 Million
Solar Energy	5 x 10	Whr/Year	32 MW SPV 4,20,000
Sq Mtrs SWH			
	4,57,000SC		
Small Hydro	10,000	MW	155 MW
Wind Energy	20,000	MW	970 MW
Energy Recovery from			
Urban & Industrial Wastes	—		4.75 MW
Ocean Thermal	50,000	MW	—
Sea Wave Power	20,000	MW	—
Tidal Power	9,000	MW	—

Source: Ministry of Non-Conventional Energy Sources, Government of India, Annual Report 1997-98.

SPV = Solar Photovoltaic. SWH = Solar Water Heater. SC = Solar Cooker

420.6 bkwh of electricity in 1997-98 comprising of thermal - 336.1 bkwh; hydro - 74.5 bkwh; and nuclear - 9.0 bkwh. Thermal plants account for 79.9 per cent; hydro electric plants for 17.7 per cent and nuclear plants for 2.4 per cent of power generation. This recorded a growth of 6.6 per cent over the previous year as shown in Table 1.3.

TABLE 1.3: Trends in Indigenous Production of Commercial Primary Energy

Year	Coal mt	Lignite mt	Crude Oil mt	Natural Gas bcum	Hydro TWh	Nuclear TWh	Wind Power TWh
1953-54	33.0	—	0.26	—	2.52	—	—
1960-61	55.67	0.05	0.45	—	7.84	—	—
1970-71	72.95	3.39	6.82	1.4	25.25	2.42	—
1980-81	114.1	4.80	10.51	2.4	46.54	3.00	—
1990-91	211.73	14.07	33.02	18.0	71.66	6.14	0.03
1996-97	288.65	22.54	33.87	22.9	68.63	9.01	0.85
1997-98	295.9	23.05	33.86	26.40	74.5	10.00	1.00

Source: Planning Commission, Government of India (1998); Indian Economic Survey 1998-99 (1999); and ONGC (1999).

Tables 1.4 and 1.5 reflect pattern of electricity consumption by different sectors and installed generation capacities.

TABLE 1.4: Pattern of Electricity Consumption (from Utilities) (1997-98)

Sector	Percentage consumption
Domestic	19.8 per cent
Commercial	6.2 per cent
Industry	37.8 per cent
Traction	2.3 per cent
Agriculture	29.6 per cent
Others	4.5 per cent

Source: Indian Economic Survey 1998-99 (1999).

TABLE 1.5: Installed Plant Capacities and Electricity Generation (March 1998)

Source	Installed Capacity	Generation
Utilities		
Hydro	21,900 MW	74.5 bkwh
Thermal	65,000 MW	336.1 bkwh
Nuclear	2,000 MW	10.0 bkwh
Sub-Total	89,100 MW	420.6 bkwh
Non-Utilities	12,900 MW	43.6 bkwh
Total	102,000 MW	464.4 bkwh

Source: Indian Economic Survey 1998-99 (1999).

Note: Figures may not add up to the total owing to rounding off.

Energy Imports

Energy imports are largely in the form of crude oil and oil products. The import/export of other commercial energy forms, viz, coal and secondary electricity, has been rather limited and restricted.

The import of crude oil and oil products in the year 1997-98 has been 34.49 mt and 19.5 mt (crude equivalent) respectively. Import of coal in the year 1997-98 has been 17 mt.

Primary Energy Supply

Total supply of energy (both of commercial and non-commercial forms excluding draught animal power) increased from 89.6 mtoe in 1953-54 to about 377 mtoe in 1996-97. Share of non-commercial fuels declined from 74 per cent in 1950-51 to about 34 per cent in 1996-97. Fuelwood continues to maintain its primacy in the non-commercial energy supply (largely in the household sector). (The break-up for 1997-98 has still to be worked out as the figures for energy production and supply are still provisional). Changes in the pattern of primary energy supply since India's planned development in 1952 is reflected in Table 1.6.

TABLE 1.6: Changes in the Pattern of Primary Energy Supply (mtoe)

Production

	1953-54	1960-61	1970-71	1980-81	1990-91	1996-97	1997-98*
Commercial Primary Energy							
Coal	23.62	35.64	36.48	56.96	94.68	122.79	125.79
Lignite	—	0.01	0.81	1.23	3.34	5.04	5.20
Crude Oil	0.19	0.46	7.01	10.79	33.92	33.87	33.86
Natural Gas	—	—	0.60	1.41	11.73	20.05	24.70
Hydro Power	0.24	0.67	2.17	4.00	6.16	5.90	6.45
Nuclear Power	—	—	0.63	0.78	1.60	2.35	2.50
Wind Power	—	—	—	—	—	0.07	0.10
Total	24.05	36.78	47.67	75.19	151.43	190.30	198.60
Net Imports	2.20	6.04	12.66	24.63	31.69	61.63	61.63
St. Changes (-)	0.24	2.87	0.69	3.80	5.37	2.38	2.38
Intl. Bunkers (-)	0.53	2.50	0.24	0.21	0.14	0.17	0.17
Total Commercial Energy Supply	25.48	39.45	59.40	95.81	177.61	249.38	257.68
Non-Commercial Primary Energy Supply	64.13	74.38	86.72	108.48	122.07	127.51	127.51
Total Primary Energy Supply	89.61	103.83	146.12	204.29	299.68	376.89	385.19

Source: Planning Commission, Government of India (1998).

* Estimated/Provisional.

Supply (also considered consumption) of non-commercial fuels in the year 1996-97 is estimated at173 mt of fuelwood, 95 mt of dung cakes and 53 mt of agro waste. Corresponding figures for 1997-98 are still being collected.

Energy Consumption

Share of commercial energy in the final energy consumption increased from 28.4 per cent to 65.9 per cent during this period, while that of non-commercial energy sources has declined from 71.6 per cent (in 1953-54) to 34.1 per cent (in 1996-97). (see Table 1.7).

TABLE 1.7: Changes in Pattern of Energy Consumption (in per cent)

Year	Main Energy Form		Break-up of Commercial Energy Consumption				
	Non-Comml.	Comml.	Coal	Oil	Natural Gas	Electricity	Total
1953-54	71.6	28.4	80.1	16.7	0.0	3.2	100.0
1960-61	65.3	34.7	75.3	19.9	0.0	4.8	100.0
1970-71	59.4	40.6	56.1	34.1	0.6	9.2	100.0
1980-81	53.1	46.9	47.9	40.3	1.1	10.7	100.0
1990-91	40.7	59.3	35.9	43.6	5.5	15.0	100.0
1996-97	34.1	65.9	29.3	46.8	6.9	17.0	100.0

Source: Planning Commission, Government of India (1998).

The shift in the relative share of final commercial energy consumption in different sectors is given in Table 1.8.

TABLE 1.8: Percentage Share in Final Commercial Energy Consumption by Different Sectors (in per cent)

Sector	1953-54	1960-61	1970-71	1980-81	1990-91	1994-95
Industry	39.9	41.5	49.6	49.6	44.7	42.1
Transport	43.7	40.2	28.2	23.5	19.8	20.8
Household	9.6	9.7	12.2	9.7	11.8	12.3
Commercial	0.5	0.3	1.2	0.9	1.1	1.2
Agriculture	1.4	1.6	2.7	6.3	8.2	9.3
Feedstocks	0.5	0.3	4.2	5.6	10.4	10.3
Non-energy use	2.9	4.7	1.5	4.0	3.1	3.0
Others	1.5	1.7	0.4	0.4	0.9	1.0
Total	100	100	100	100	100	100

Source: Planning Commission, Government of India (1998).

The total consumption of petroleum and petroleum products during 1997-98 is reported at 84.5 mt (80.1 mt from the public sector units and 4.4 mt imported by the private parties).

Energy Development Programme upto 2002

Coal

The Ninth Five Year Plan, for the period 1997-2002 is in the state of finalisation. However, indications are that the programme for coal sector includes mine development to increase the annual production capacity of coal from a level of 289 mt in 1997 to 371 mt in the year 2001-02.

Besides the increase in physical output, the programme for five years (1997-2002) includes: intensified detailed exploration efforts; proving additional coal resources; improving coal quality and consistency through coal washeries for coking coal; and beneficiation of non-coking coal being supplied to power plants situated at distances beyond 1000 kilometers. Other measures proposed include; environmental measures; improved coordination for coal transportation; and enhanced R&D thrust in the areas of productivity, mine safety and coal beneficiation.

Oil and Natural Gas

The programme for the oil sector includes augmenting the domestic crude production to reach 37 mt and that for gas production to reach 31 bcum in the year 2002. Oil demand is expected to reach 113 mt in the year 2001-02. The annual import of crude oil and oil products is expected to reach 76 mtoe in the year 2001-02 making import dependence at 70 per cent.

Following areas are being given attention :

(i) Improving the reservoir management and enhancing oil recovery;
(ii) Rationalisation of pricing structure for POL products in the domestic market in order to reduce demand;
(iii) Creation of additional refining capacities and marketing and distribution facilities;
(iv) Exploring the possibility of substitution of POL products by LNG.

Electric Power

The gross energy generation from utilities in the year 1997-98 has been 420 TWh, indicating an average annual growth of 6.6 per cent over the past five years. The programme for the five years (1997-2002) involves addition of over 40,000 MW of capacity with expansion of associated transmission systems and distribution networks.

Power development programme includes: participation in the capacity addition programme through private entrepreneurial support; emphasis on renovation and modernisation of existing facilities; rationalisation of mismatch in the transmission and distribution support; reduction in the regional disparities in the electricity availability; reduction in the average gestation time per megawatt of addition to installed generating capacity; better fuel management with targeted reduction in the specific fuel consumption levels; and encouragement to captive power generation, including co-generation.

Nuclear Power

Nuclear power development programme for the next five years includes; modest addition to the physical capabilities in the nuclear power generation; thrust on performance of nuclear power stations; and upgradation of capabilities for higher capacity addition programme based on indigenous capability.

Non-conventional Energy Sources

Importance of renewable energy has significantly increased in recent years because of growing environmental concerns associated with fossil fuel based energy and the urgency of promoting sustainable energy development. Renewable energy also has a particularly critical role in meeting the growing energy needs in the rural areas. Furthermore, the large scale development and utilisation of renewable energy sources would also lead to the strengthening of energy security by reducing the import bill on fossil fuels.

Energy Policy

Strategic Elements of Energy Policy

Energy strategies adopted from time to time are preceded by detailed study/studies by expert groups. The first such study was carried out in 1961, under the aegis of the "Energy Survey of India Committee", with a comprehensive survey of energy resources of the country which enabled formulation of a pragmatic energy strategy during the sixties. During the period between the two oil shocks (1973-79), the energy situation was reviewed twice—once under the aegis of the Fuel Policy Committee in 1974, and again under the aegis of the Energy Policy Committee in 1979. In 1976, A comprehensive Energy Policy was formulated which, underlined *inter alia* :

- — importance of coal as the principal source of energy;
- — need for reducing reliance on imported oil;
- — need for substitution of oil by any and all other forms of energy, wherever technically and economically feasible;
- — need for developing new and renewable energy sources with appropriate levels of efficiency in use together with progressive improvements in the efficiency of their use;
- — improving the efficiency of use in the existing installations through technology and management back-up and renovation programmes, efficiency improvements in the production systems and reduction in transformation losses, including those in traditional forms of energy resources;
- — adopting energy efficient technologies in the supply and enduse systems;
- — in-building and internalising standards for environmental efficiency in the energy production and supply processes and systems;
- — rationalising energy pricing in order that real costs are reflected in the revenue making processes;
- — strengthening and reforming the institutional structures in the energy sector in order to promote the economic policies in the new and adapted atmosphere;
- — promoting R&D in the energy sector to integrate economically the elements of environmental benignness, cost sustainability in the commercialisation of the energy technologies, including those meant for substitution of depletable energy resources by renewable resources and improved utility and convenience in the secondary energy use.

Energy Sector Investment

Investments in Different Five Year Plans

Energy sector development since independence has been based on government support and public fund allocation. Power (electricity) sector was taken over by the government in view of its direct impact on industrial and agricultural growth. Subsequently, the coal and the oil sectors were also nationalised to meet the social and economic objectives.

There has been priority channelisation of public investments to the energy sector programmes. Every Five Year Plan allocated upto 30 per cent of its funds for the energy sector; electricity sector receiving allocation support to the extent of 20 per cent of the outlay. Average annual investment in the energy sector was Rs. 772 million per year during the five year period 1951-56 which went up to Rs. 230 billion per year during the period 1992-97. The investment trends/plan expenditure during the five year periods from 1951-56 to 1992-97, in absolute figures and the share of energy sector investments in the overall annual public sector investments/outlays is given in the Table 1.9.

TABLE 1.9: Plan Expenditure and the Share of Energy Sector

Plan	Period	Plan	Share of Energy Sector (%)				
Expend.		(Rs. billion)	Power	Oil/Gas	Coal	NRSE	Total
First	1951-56	19.6	—	—	—	—	19.7
Second	1956-61	46.7	9.7	0.8	1.9	—	12.4
Third	1961-66	85.8	14.6	2.6	1.3	—	18.5
Fourth	1969-74	157.8	18.6	1.9	0.7	—	21.2
Fifth	1974-79	394.3	18.7	3.6	2.9	—	25.2
Sixth	1980-85	1092.9	16.7	7.8	3.5	0.1	28.1
Seventh	1985-90	2202.2	17.4	7.3	3.2	0.3	28.2
Eighth*	1992-97	4341.0	18.4	5.5	2.4	0.2	26.5

Source: Planning Commission, Government of India, 1998.

* Outlay

Energy Sector Investment in Post 1992 Era

After the liberalisation (in 1991) and with the opening up of economy to private sector, there has been reduction in the public investment to energy sub-sectors in the hope that the private sector will provide the necessary support. Private sector investment was, however, enough to meet the additional financial needs of the power and coal sub-sectors. As a consequence, during the five year period covering 1992-97, the investment in the power sector has declined to 2 per cent of GDP as against 2.4 per cent during the Sixth Plan (1980-85). Similarly, public investment in coal declined from 0.5 per cent of GDP in the Sixth Plan (1980-85) to 0.3 per cent of GDP in post 1992. On the other hand, investments in oil and gas exploration, oil refining and gas supply increased from 0.7 per cent of GDP in the Sixth Plan (1980-85) to 1.2 per cent of GDP in post 1992.

Over 90 per cent of investment in the energy sector, till the liberalisation, has been under public sector ownership.

Inspite of motivation of the private sector, the government will continue to play an important role in the energy sector projects. In that context, following issues deserve attention :

(i) Financial guarantee to domestic Development Financial Institutions (DFI) through the provision of sovereign guarantees, equity capital and preferred tax provision.
(ii) Promotion of growth of Domestic Institutional Investors by the government by encouraging higher growth rate of institutional assets.
(iii) Promotion of various other options such as : shift of energy investments from capital to current account by importing energy goods and services; shifting investments to non-energy sectors and privatisation.
(iv) Development assistance and soft term loans from international funding agencies.
(v) Involvement of multinational companies in joint ventures.
(vi) Participation by public and private companies and employees.
(vii) Leasing of energy equipment.
(viii) Institutional development.

Energy Sector Reforms

Energy Sector Reforms were introduced to promote foreign and private participation in the following categories :

(i) Setting up of Electric Generation Stations.
(ii) Coal Mine Development.
(iii) Exploration, Proving and Development of Oil & Natural Gas wells.

Functional and structural reforms were launched in all energy sub-sectors. The major sub-sector to attract attention of the developer and private investor was the power sector. After the initial teething problems, the Government of India could streamline the procedures and restore investor confidence. The 17th Congress of the World Energy Council, at Houston in September, 1998, bears testimony to the fact that ever increasing number of foreign developers are generating understanding of the situation in India and hoping to launch their programmes at a fast pace. A special attraction at the Congress was the visitors at the India Pavilion where a large number of foreign oil companies and investors were seeking clarification for potential investment, also in oil sector. Thus, India is becoming an attractive destination for foreign investors, both for power and oil sectors. Following specific functional and structural reforms in the energy sector will be of greater interest to indicate as to how the energy sector is moving in India.

Power Sector Reforms

Functional Reforms (Investment Promotion)

Reforms in the power sector were initiated in 1991. In order to make the climate for investment congenial to the entrepreneurial measures for deregulation and devolution of power to utilities, direct negotiation with the entrepreneurs has been encouraged. Regulations relating to fuel supply have been liberalised so as to attract a variety of fuels to become supportive to power generation based on economy of supply. Legislative measures were effected to support congeniality in the commercial climate in order to make tariffs acceptable to the consumer and encouraging the entrepreneurs. The clearance procedures have been rationalised. Time for techno-economic clearance by CEA has been reduced to 5-6 months. Project clearance based on competitive bidding is made mandatory and the methods for clearance are being rationalised.

Structural Reforms

In structural terms, power sector utilities are endeavouring to unbundle the existing vertically integrated monopoly and separate generation, transmission and distribution activities for rationalising their management. Whereas efforts are made to maintain conceptual uniformity in the unbundling, the multi-ownership for generation and distribution and continuation of government monopoly in the transmission would be the near term approach in the initial stage of reforms. Utilities are currently debating the possible structures. The existing system (State Electricity Board structure) in some utilities may continue in the transition period with functional and managerial improvements (as a few utilities have done creditably well even in the existing dispensation of vertically integrated structures) may be allowed. Orissa State Electricity Board was the first to introduce reforms and set up three different organisations to cater to generation, transmission and distribution. Andhra Pradesh has also successfully introduced a similar transformation. Haryana has planned similar action.

Setting up Electricity Regulatory Commissions

In order to provide a congenial and confidence building climate and to promote healthy competition and reasonable and transparent tariff making arrangements, the Central Electricity Regulatory Commission (CERC) has been set up in 1998 at the central level. CERC has been charged with the following responsibilities :

(i) to regulate the tariff of generating companies owned or controlled by the central government;
(ii) to regulate the inter-state transmission including tariff of the transmission entities;
(iii) to regulate the inter-state bulk sale of power;
(iv) to aid and advise the central government in formulation of the tariff policy etc.

The process of unbundling of generation, transmission and distribution, establishment of regulatory mechanisms and readjustments in the existing structures are at different stages of implementation in different federal utility systems. In Orissa, the State Electricity Regulatory Commission has since been set up. Regulatory systems would be on anvil in Rajasthan and Gujarat . On the other hand, in states like Andhra Pradesh, Maharashtra and Tamil Nadu, attention is being paid for readjusting the existing structures. In other states, the choice of the unbundled systems is at different stages of discussion and debate. Other areas that would come under the purview of the commissions (at the state level) will be regulation of the working of licensing, protection of the interest of the consumers, ensuring reasonability in the operational efficiency and pricing of electricity.

Recent Legislative and Policy Measures

The following legislative and policy measures have been initiated recently

Legislative Measures

- — Electricity Laws Amendment (to facilitate transmission).
- — Electricity Regulatory Commissions Act enactment.

Policy Measures

- — New Liquid Fuel Policy announced.
- — Energy Conservation Bill to ensure optimum utilisation of energy to be finalised shortly.
- — Central Electricity Regulatory Commission established.
- — Revised Counter Guarantee procedure finalised.
- — Techno Economic Clearance (TEC) given for 48 projects.
- — 4 per cent interest subsidy on PFC loans under accelerated generation and supply, for Ninth Plan period, for expeditious completion of renovation, transmission and generation projects of State and Central power utilities.
- — Additional 1 per cent rebate on PFC loans to States setting up Regulatory Commissions.
- — Special Action Plan and Package for North East States including additional 1 per cent subsidy on PFC loans.
- — Public investments in power sector raised to Rs. 9,950 crores in 1998-99, as against Rs. 8,188 crores in 1997-98
- — State governments empowered to clear projects facilitating increased electricity generation.
- — Efficiency of existing power utilities improved, facilitating increased electricity generation.
- — Comprehensive Hydro Policy finalised to facilitate more hydro power projects.
- — Introduction of Availability Based Tariff for rationalising power tariff structure and stabilising availability of power.
- — Policy for facilitating Mega Projects formulated and Foundation Stone for 2 Mega Projects laid in March 1999.

Some provisions of the budget proposals for the year 1999-2000 have a direct bearing on power sector growth. These are as follows:

- — The facility of a five-year tax holiday and deduction of 30 per cent of profits in the subsequent five years available to companies engaged in generation and distribution of power has now been extended to companies setting up transmission lines on or after April 1, 1999. This will provide an impetus to reforms of state electricity boards, which are unbundling generation, transmission and distribution. It will also encourage greater private sector participation in transmission.

— Unlike mega power projects which are entitled to zero customs duty on imports, IPPs will pay a basic customs duty of 5 per cent, a countervailing duty of 16 per cent and a surcharge of 10 per cent.
— Amendment of Section 80-A of Income Tax Act would be especially useful for power projects that are likely to have revenue later in the project life cycle.
— Import of equipment for power generation projects will see a net excise decrease of 1 per cent while imports of equipment for power transmission projects of 65 kV and above will have to pay an additional 9 per cent.
— Reduction in customs duty for R&M of power generation plants excluding captive plants.
— The basic customs duty on plants increased from 22 per cent to 25 per cent for captive power.

Oil Sector Reforms

Functional Reforms (Investment Promotion)

Reforms in the oil sector include allowance to privàte participation in exploration, production and marketing activities of the crude; and marketing of specific oil products in selected areas. Initiatives are taken for rationalisation of pricing policy for the individual products for greater reflection of the real costs on the prices. Joint ventures in the exploration and drilling of new wells will be encouraged. Deregulation of the natural gas sector to encourage private participation in exploration, production and transportation of the natural gas within the country is on the anvil. This may pave the way for independent regulation for gas transportation and distribution.

Structural Reforms

In order to make the functional reforms workable and arrange a congenial atmosphere for private participation, regulatory mechanisms through a coordinated approach at the national level has been initiated with the establishment of a Directorate General of Hydrocarbons (DGH) as a monitoring entity for regulation. In the downstream, the Oil Coordination Committee (OCC) was set up to provide technical and operational support in the implementation of regulatory policies. Presently, the OCC is operating as a national downstream "corporate planning-cum-supply planning" entity. The role of OCC is likely to be altered in order to make the OCC a monitoring agency.

Following advancements have been reported in the petroleum sector, since April 1, 1998.

— Phased dismantling of Administered Pricing Mechanism commenced from 1st April, 1998. Prices of all products except MS, HSD, ATF, LPG (domestic), SKO (PDS) decontrolled.

— Price of diesel linked to import parity price. Price of HSD reduced substantially thrice in April, May, 1998 and January, 1999.

— Refining Sector delicensed. 5 Year Tax Holiday granted for new refineries set up after 1st October, 1998.

— New refining capacity added. Two new refineries at Panipat and Numaligarh completed. Foundation stone of Bhatinda (Punjab) Refinery of 9 million tonnes capacity laid in November, 1998. Government approval accorded to 9 MMTPA Refinery at Paradeep (Orissa). Refinery at Nagapattinam (Tamil Nadu) of 9 MMTPA proposed by IOC.

— A progressive and attractive New Exploration Licensing Policy (NELP) finalised and announced.

— Bids invited for 48 exploration Blocks including 12 deep sea offshore Blocks which hold large hydrocarbon potential, under NELP.

— Petronet LNG, a new company formed and action initiated to set up LNG import terminals at Dahej,Gujarat (5 MMTPA); and Kochi, Kerala (2.5 MMTPA) to meet the growing demands of natural gas for power and fertilizer sectors.

— ONGC Videsh Ltd initiated action to acquire stakes in oil production in Iraq and Russia.

— To contain the environmental pollution, only unleaded petrol (ULP) supplied in Delhi with effect from 1st September, 1998. ULP introduced from June, 1998 in all other Metropolitan cities, State Capitals and major towns.

— 3 New Product Pipelines (Vadinar - Kandla, Cochin - Karur and Mangalore - Bangalore) under implementation by the newly set up Petronet India Ltd in the direction of creating a national pipeline grid.

— Functional autonomy and Navaratna Status for National Oil majors viz, IOC, BPC, HPC, GAIL and ONGC through reconstitution of Boards by including professional non-official Directors.

The budget proposals for 1999-2000 have mentioned specifically of the following:

— Basic customs duty on the following has been reduced:
  -crude petroleum from 22 per cent to 20 per cent
  -high speed diesel from 32 per cent to 30 per cent
  -LNG from 12 per cent to 5 per cent
— Customs duty on naphtha has been hiked from 0 per cent to 5 per cent. This reflects the desire of the government to discourage naphtha-based power plants and push LNG as a fuel.
— Across the board 10 per cent surcharge on customs duty does not apply to crude and petroleum products.

Coal Sector Reforms

Functional Reforms (Investment Promotion)

The coal sector reforms include offers for participation by private sector for mining in potential blocks for augmenting the annual coal production. Legislative measures to effect deregulation in the coal sector are being completed.

The deregulation includes :

(i) freedom to consumers to choose the coal supply sources;
(ii) negotiation of quantity and price in a market-driven atmosphere and competitive climate;
(iii) measures to diversify the choice for the consumer in terms of resilience in the indigenous supply; and
(iv) import of coal.

Competitive bidding for offering new mine development programmes in the private sector would be introduced. Setting up of the Regulatory Commissions and adaptations in the institutional structure, encouragement of joint ventures between the existing public sector core companies and private sector will also be encouraged.

In order to make the coal industry financially viable and self-supporting, following measures have been initiated :

(i) Rationalisation of the manpower.
(ii) Decontrol of prices of all non-coking superior grade coals and coking coals.
(iii) Closure of uneconomical mines.
(iv) Capital restructuring packages for collieries.
(v) Reduction of import duty on equipment and spares.
(vi) Parity for distribution of imported coal and indigenous coal.
(vii) Implementation of cash and carry system for coal supplies.

Structural Reforms

The structural reforms aimed at encouraging the private participation in coal mining and builtup of washeries include the following :

(i) Provision for captive coal mining by consumers, and permission to sell coal from April 1, 2000.
(ii) Construction of washeries on Build-Own-Operate (BOO) basis.
(iii) Setting up of washeries by power stations.
(iv) Setting up of washeries by private agencies as agents to the power stations.

A significant decision has been taken by the Cabinet to allow coal trading by private companies from captive mines allocated for power projects. This decision has to be translated into a Bill and presented to the Parliament. The government has assured all clearances before April 1, 2000. The private companies will be then free to undertake coal trading.

The budget proposal for 1999-2000 raised effective duty on imported coal from 15 per cent to 16.5 per cent.

Energy Perspective 2020

Issues in Perspective

The Indian Member Committee of the World Energy Council, under

IE(I), have focused on projections to year 2020 as the 25 year time frame can be understood closely and forecast made comprehensively. The conclusions of the 17th Congress of the World Energy Council also emphasised on suitability of present technologies for application upto 2020, thereafter the emerging technologies will take over.

India's population may exceed 1.3 billion by the end of 2019-20 involving urban/rural population ratio at 43 per cent/57 per cent in that year. Rapid urbanisation and literacy improvement in the rural areas, are likely to be accompanied by high rates of economic growth in the range of 6-7 per cent. That will bring about an appreciable increase in the per capita income and hence change the life styles. That will influence the energy demand. There is a growing perception that environment protection and conservation would have to be adequately internalised in the growth of the energy sector. The key issues involving the energy demand implications therefore include :

(i) rise in population;
(ii) sustainability of economic growth;
(iii) accessibility to adequate commercial energy supply;
(iv) accessibility to financial resources needed to support expansion of the energy supply systems; and
(v) rational price regime.

Energy Demand 2020

A few model studies are made by experts in the recent past for projecting the energy demands in the global perspective for the period upto 2020. The latest study by E. A. S. Sarma, J. N. Maggo and A. S. Sachdeva indicates the overall energy demand perspectives for three different scenarios as follows :

(i) a reference scenario in which the business as usual (BAU) would be allowed to continue,
(ii) an efficiency oriented scenario (EFF) considering the scope for reduction in the energy intensity based on induced efficiency through technology and management adaptations, and
(iii) an environmentally constrained scenario (ENV) considering the implications for emphasis on minimisation of adverse environmental impacts in augmenting the energy supply system.

Projections were made under the three different scenarios for the different sectors of consumption and for three terminal years viz. 1999, 2009 and 2019. Projections made under the above for the "business as usual" scenario indicate the total demand requirement by 2019 would work out to 39107 peta joules (934.0 mtoe) as compared to 9727 peta joules (232.3 mtoe) in 1989. However, looking from the point of view of environmental sustainability and huge investment involvement, energy conservation and energy efficiency programmes would have to be focussed. Their successful implementation can reduce the demand substantially.

In order to meet the level of final energy demand projected in the BAU Scenario, the installed capacity in power generation will need to increase to nearly 3,85,000 MW in the year 2020. The coal and oil requirement will correspondingly be 1,400 mt and 400 mt respectively.

Expert consultations at the Asia Energy Vision 2020, organised under the auspices of the Indian Member Committee of the World Energy Council, in New Delhi in November 1996 agreed on these energy demand projection of India. The details are given in Table 1.10.

TABLE 1.10 : Future Demand for Commercial Energy in India

Source	Unit	1991-92	1996-97	2009-10	2020-21*
1. Electricity	TWH	231	336	725	1350 (1745)**
2. Coal	mt	229	311	690@	1400
3. Petroleum Products	mt	57	81.2	165	335
4. Natural Gas	bcum	18.6	30.2	65	130

Source: Planning Commission, Government of India - September 1995 & September, 1996 and India's National Paper on Energy Scene 1998 presented at the 17th Congress of the World Energy Council, at Houston in September 1998.

* Projections to 2020-21 made on the basis of 'business as usual' - J.N.Maggo - R&D Needs for Energy Sector - September, 1996.

** Electricity generation capacity of 1745 TWH will be required to take care of T&D losses and Plant Load Factor. Corresponding installed generation capacity will be 385,000MW at present day PLF and T&D loss level (1998).

@ R.K.Sachdev - International Conference on Coal - June 1996.

Projection of requirement of Coal in 2020-21 has also been revised in 1998.

REFERENCES

Annual Report 1997-98, Ministry of Power, Government of India.

Annual Report, Ministry of Coal, Government of India.

Annual Report, Ministry of Petroleum & Natural Gas, Government of India.

Annual Report, Ministry of Non-Conventional Energy Sources, 1997-98, Government of India.

Economic Survey of India 1999, Government of India.

Gangawar, Plenary Address-Petrotech 99 Conference, New Delhi, January 9, 1999.

National Paper, India's Energy Scenario 1998, Presented at the 17th Congress of WEC at Houston, September 13-18, 1998.

Ramamurthy, K. Theme Address-Petrotech 99 Conference, New Delhi, January 9, 1999.

Sarma, EAS, *et al.*, *Projections for Energy Consumption in India Study,* Planning Commission, 1997.

Vajpayee, Atal Behari, the Prime Minister of India, Inaugural Address-Petrotech 99 Conference, New Delhi, January 9, 1999.

2

TECHNOLOGY DEVELOPMENT FOR ENERGY SYSTEMS

Introduction

Global energy supply pattern is assuming new dimension. Energy availability is projected to meet all the needs of global populations. Similarly the finance for inducting new energy supply and conversion systems are projected to be sufficient. While these global figures are satisfying they are not sufficient to justify situation at every place of the globe. The reason lies in identifying the financial capabilities of the needy and the location of supply options. Added to this is the dimension of environment control. Industrialized countries have the means and experience to undertake suitable measures, but most of the developing countries do not have that kind of capability. While the energy developers are discussing about the market distortions due to subsidies and other financial support in certain parts of the globe, the major problem now experienced by the World Bank and other donors is 'market exclusion' of large populations.

17th Congress of the World Energy Council: Observations

Primary Concern - Serve the Poor without Access to Commercial Energy Sources

17th Congress of the World energy Council was organised at Houston during September 13-18, 1999. The Congress, while focusing on the theme "Energy and Technology; Sustaining World Development into the Next Millennium" identified that 2 billion people in the developing

---

Paper presented at the Energy Summit , organised by the Confederation of Indian Industry at Chennai on December 13, 1998.

world have no access to commercial forms of energy, which has also contributed to their underdevelopment. It also realised that the market pricing of energy only make an investment attractive if there is a market that can pay the price; the poor in the developing countries do not yet constitute such a market. Any such development which does not serve the poor is really not conducive to socio-economic development of the globe. However, it also realised that the pace at which the poor become empowered energy consumers will depend more on the success of policies to reduce poverty than on energy policies.

Accelerated Development and Use of Renewables Necessary

The Congress identified that accelerated development and use of appropriate renewable energy resources must be given high priority as a means of supplying commercial energy services to people for the first time. Active solar, biomass, wind and mini-hydro development schemes were endorsed. WEC decided to encourage new partnerships and new economic systems to sustain long-term investment programmes designed to bring energy to rural areas.

Emerging market conditions have also forced energy companies to look into potential of renewable energy use. Major oil companies, in Europe and USA, are already backing up renewable energy units. Major energy companies, in India, like ONGC and NTPC have also initiated programmes to support renewables.

Nature of Energy Business will Change

The participants focused their attention on the direction that the energy markets could take in the next 50 years. It was recognized that the old certainties and paradigms are giving way to new markets, challenges, organisations and ways of doing things. The major petroleum industries or the major current players expressed their concern that in over a decade a large number of major energy players may be reduced to producers and sellers but not the leaders in the energy sector. Future lies in the hands of such industries who have linked themselves with information technology and are willing to change with time. The expectation was that a large number of smaller players, who build up their outfit with information technology back-up will control energy sector. Mega mergers in the petroleum sector have been witnessed in the global markets; and collaborative programmes amongst the leading players in the Indian Petroleum sector have also been witnessed.

On the other hand the public utilities in the developing countries are unbundling vertically - splitting into generation, transmission and distribution entities. The utilities are considering the possibility of integration at the application or use end by providing combined heat and power application in the developed countries.

Technologies Important but not Sufficient

The deliberations identified that the technologies will play an important role; but, in themselves will not be sufficient to fix all problems. Even development of technologies and their implementation will call for major policy and financial support from the government. Therefore, institutional and policy changes will be required. Options will have to be made that are attractive for international funding at right places. Public awareness and public attitudes and behaviour will need a major shift. Public curiosity will not be satisfied by information dissemination alone but they will be keen to receive answers for the queries that come up from time to time. The approach will shift from energy being a 'commodity industry' to a 'service industry'.

WEC Study on Global Energy Perspectives

Every Congress of the WEC has a major Report as the document of the Congress that gives future perspectives. A continuity in operations has been maintained since 1993. IIASA and WEC have jointly run a Study now spanning six years. This Study has helped to identify the future indicators of energy sector market development. These are by no means to be taken as forecast but only as indicators. They give a better idea about the kind of actions desirable from member-countries to ensure smooth energy supply.

Energy Demand and Supply Perspectives

The Study projects as follows:

The Study - Global Energy Perspectives presents three cases of future development sub-divided into six alternative scenarios. The principle focus is on the period between 2020 and 2050, but some results are also presented out to 2100. In brief, Case-A presents a future of impressive technological improvements and consequent high economic growth. Case-B describes a future with less ambitious, though perhaps more realistic, technological improvements, and consequently more intermediate economic growth. Case-C presents a 'Rich and Green Future'. It includes both substantial technological progress and unprecedented international cooperation centered explicitly, on environmental protection and international equity.

All the three cases provide for substantial social and economic development, particularly in the developing world. They provide for improved energy efficiencies and environmental compatibility, and thus, for associated growth in both the quantity and quality of energy services. Across all three cases, the structure of final energy develops in the same way and the energy intensities improved steadily. To facilitate comparisons among the three cases, all share the same central demographic baseline assumption in which global population grows to 10 billion to 2050 and to nearly 12 billion by 2100. All have been checked for internal consistency with the aid of formal models and through a lengthy regional expert review model.

In Case-A, global primary energy consumption during 2050 and 2100 will be 25 Gtoe and 45 Gtoe respectively (compared to 9 Gtoe in 1990). In Case-B the figures will be 20 Gtoe and 35 Gtoe and in Case-C the figures will be 14 Gtoe and 21 Gtoe.

It is easier to anticipate the forms in which energy will be demanded by consumers in the future than to estimate the absolute level of energy demand, or which energy sources will supply that demand. With increasing per capita incomes around the world, people will demand higher levels of more efficient, cleaner, and environmentally less obtrusive energy services.

However, the question of what kind of companies will supply energy services, and how, is wide open.

Three major issues for decision on energy options will be as follows:

— Which energy sources will best match consumer's preference for flexible, more convenient, and cleaner energy services?
— Which companies will have made the investments in Research Development and Demonstration (RD&D) that will give them a technological edge?
— Which companies will have refocused their business from providing just tonnes of coal or kilowatt-hours of electricity, to providing convenient and clean energy services to consumers?

The answers to those questions will be determined between now and 2020. Because of the long life times of power plants, refineries, and

TABLE 2.1 Summary of the three cases in 2050 and 2100 compared with 1990

	Case		
	High growth	Middle course	Ecologically driven
Population, billion			
1990	5.3	5.3	5.3
2050	10.1	10.1	10.1
2100	11.7	11.7	11.7
GWP, trillion US (1990) $			
1990	20	20	20
2050	100	75	75
2100	300	200	220
Global Primary energy intensity improvement, percent per year	Medium	Low	High
1990 to 2050	-0.9	-0.8	-1.4
1990 to 2100	-1.0	-0.8	-1.4
Primary energy demand, Gtoe			
1990	9	9	9
2050	25	25	25
2100	45	35	21
Resource availability			
Fossil	High	Medium	Low
Non-fossil	High	Medium	High
Technology costs			
Fossil	Low	Medium	High
Non-Fossil	Low	Medium	Low
Technology dynamics			
Fossil	High	Medium	Medium
Non-Fossil	High	Medium	High
Environmental taxes	No	No	Yes
CO_2 emission constraint	No	No	Yes
Net carbon emissions, GtC			
1990	6	6	6
2050	1-15	10	5
2100	6-20	11	2
Number of scenarios	3	1	2

Abbreviations: GWP = gross world product; Gtoe = gigatonnes oil equivalent; CO_2 = Carbon dioxide; GtC = gigatonnes of carbon

Source: WEC/IIASA Study on Global Energy Perspectives, 1996.

other energy investments, there is not a sufficient turnover of such facilities to reveal large differences in our scenarios prior to 2020, but the seeds of the post-2020 world will have been sown by then. The choice of the world's post-2020 energy systems may be wide open now. It will be a lot narrower by 2020.

Specific Observations of the Study

Following major observations were made under this study

(i) World energy needs will increase.
(ii) Energy intensities will improve significantly.
(iii) Resource availability will not be a major global constraint.
(iv) Quality of energy services and forms will increasingly shape future energy systems.
(v) Energy end-use patterns will converge, even as energy supply structures diverge.
(vi) Technological change will be critical for future energy systems.
(vii) Rates of change in global energy systems will remain slow.
(viii) Interconnectivity will enhance cooperation, systems flexibility, and resilience.
(ix) Capital requirements will present major challenges for all energy strategies.
(x) Regional differences will persist in global energy systems.
(xi) Local environmental impacts will take precedence over global change.
(xii) Decarbonization will improve the environment at local, regional, and global levels.

Summary of the Study are given as Annexure-I and Conclusions and Recommendations are given as Annexure-II at the end of the book.

WEC-IIASA Observation on Technology Development

The WEC-IIASA Report has observed as follows:

Technological change drives productivity growth and economic development. Progress has a price. RD&D of new energy technologies and the accumulation of experience in niche markets requires up-front expenditure of money and efforts. These are increasingly viewed as too high a price to pay in liberalised markets where the maximisation of short term shareholder value generally takes precedence over longer term socio-economic development and environmental protection. Yet it is the RD&D investments of the next few decades, that will shape the technology options available after 2020. A robust hedging strategy will have to be evolved to focus on generic technologies at the interface between energy supply and end use, including gas turbines, fuel cells, and photovoltaic. These could become as important as today's gasoline engines, electric motors and micro-chips.

Innovation and technology diffusion require both that opportunities are perceived and the entrepreneurial spirit exists to pursue them. Long term scenarios cannot forecast future technological winners, but they can indicate areas of technological opportunity.

Across the wide variation impossible energy developments depicted in different scenarios, the importance of energy infrastructure grows persistently The market for power sector technologies also grow substantially, with a wide spread between the maximum and minimum scenarios. Part of this spread relates to uncertainty about demand growth, but part of the spread arises from energy end use innovations in the form of new, onsite, decentralised electricity generation technologies, such as photovoltaic and fuel cells. The potential for end use technologies in the long term outgrows that of the power sector. The most important customers for energy technologies could no longer be a limited number of utility managers but rather millions of energy consumers worldwide. Synfuel also emerges in the long term as a major technology market. An orderly transition away from conventional oil and gas translates into large technology markets for syn-liquids, syn-gas, and in the long term hydrogen produced from both fossil fuel (coal and natural gas) and renewables (biomass). By the end of 21st century, the synfuel markets could be at least 4 Gtoe per year, comparable to the current oil market.

The scenario allows two additional, more specific conclusions about investment strategies:

The first, is the importance of improving and expanding infrastructures for clean grid dependent energy forms. For energy infrastructures, the back-bone of the energy system, the impact of technological change can be particularly slow because of the slow rate of capital turnover. Benefits also tend to be long term and include significant positive spin-off effects, or externalities, for which a private sector cannot profit directly. On the other hand such infrastructures have the advantage that their value is largely independent of which energy sources dominate future energy systems - the electricity grid can transmit electricity from coal as easily as electricity for renewables or nuclear power. Gas pipeline can transmit natural gas, syngas, and in the longer term also hydrogen, e.g. in the form of hythane. But because short term incentives for such needed infrastructural investments are low, these investments are an area where government intervention is appropriate and public - private partnerships desirable. From a long term global perspective, the improvement and expansion of energy grids have to be extended on Eurasia and South Asia.

The second, more specific conclusion concerns the importance of end use technologies. Although the traditional focus of energy technology policy has been on energy supplies, a clear result of this study is that long term improvements in energy system costs, efficiency and environmental impact will increasingly be determined by technological innovation at the level of energy end use. Generic technologies that are at the interface between energy supply and end use (gas turbines, fuel cells and photovoltaic) are there for a robust bet for enhanced RD&D and diffusion efforts, and could become as important as today's internal combustion engines, electric motors and micro chips. Over the long term much of today's traditional, centralised energy conversion may even shift toward the end user. For example, hydrogen powered fuel cell vehicles could, when parked, be a source of clean residential electricity. Instead of moving electrons, hydrogen molecules could be moved. In this case the importance of the energy infrastructure does not diminish, but instead grows.

The conclusion that the point of final energy use is where we expect far-reaching technological improvements to occur has two additional implications. First, it weakens the argument for extensive RD&D investments in large, sophisticated, lumpy, and inflexible technologies such as fusion power and centralized solar thermal power plants. Improvements in end-use technologies, where millions, rather than hundreds, of units are produced and used, are more amenable to standardization, modularization, mass production, and hence exploitation of learning-curve effects. Second, institutional arrangements that govern final energy use and supply are critical. Deregulation and liberalization of electricity markets can create incentives in this direction as service packages are tailored to various consumer preferences and especially as traditional consumers can sell electricity back to the grid. But there are also concerns that liberalization will discourage longer-term RD&D investments by emphasising short-term profits.

The fact that increased attention should be devoted to technological innovation focused on energy grids and energy end use does not in any way mean that the importance of primary energy and centralized conversion energy technologies will diminish. Fossil fuels, particularly natural gas, are central to energy supply during the next few decades in all scenarios, and the more efficiently and cleanly we can use them the better. In the short term, the message of the scenarios is that supply-related RD&D should be focused on developing clean fossil technologies, increasing efficiencies, and improving geological understanding of conventional and unconventional resources, particularly gas and oil. For the long term, RD&D on energy supplies should anticipate the grand transition from fossil to non-fossil energy sources. For example, experience with synthetic fuels from fossil resources will speed up the eventual diffusion of synthetic fuels from non-fossil resources. A range of non-fossil technologies should be explored along with the new and inventive conversion cycles and novel combinations of energy sources and fuels. The emphasis should be on multipurpose technologies that can be adapted to different energy sources and energy carriers. The objective is to reduce the risk of becoming locked-in to technologies and energy sources that turn out to be less advantageous as uncertainties about the future resolve with time.

Finally, the consistent emphasis on clean forms of final energy suggests that in the long term meeting even high levels of end-use energy demand can be consistent with regional and global environmental protection. For the near term it is important to accelerate the diffusion of existing environmental technologies - scrubbers, unleaded gasoline, and better stoves in rural areas of developing countries. But for the benefit of the longer term, it is important that near-term solutions do not lock in existing fossil-based technologies and slow the introduction of cleaner fossil and non-fossil alternatives. The challenge is to speed the diffusion of current clean technologies while increasing the flexibility of the energy system to explore, incorporate, and profit from yet cleaner and more efficient alternatives.

Options for Future Technology Development

A look at the history of energy substitution over the past several centuries indicate that it has always been a technological innovation, and not a pervasive shortage of wood, coal or petroleum products, which has led to the introduction of new energy sources.

Therefore, promotion of innovation, including fundamental R&D will likely to be foremost driver of technological change. With the advent of competition in energy industry, technology development and deployment takes on a new meaning. Movement away from centralised management and coordination, to profit oriented specialist along the energy supply chain implies that these entities will better manage their operating procedures and hopefully improve or re-define the value of their products and services.

Recent Trends in Energy Technology RD&D

IEA made an assessment of impact of technology and funding using data from its member countries.

Following observations have been made:

(i) Overall levels of public spending on energy RD&D are declining in the industrialised countries.

(ii) The composition of the remaining budgets has changed. Government budgets for fossil and nuclear energy technologies have declined significantly in some countries, while support for renewable energy has increased recently and energy conservation budgets have been increasing since 1985.

(iii) In the context of climate change policy, governments are examining possible increases in research on climate-friendly energy technologies.

(iv) Spending on energy RD&D may also be declining in the private sector. Private RD&D spending in the energy sector, as a percentage of yearly turnover, is now lower than in high-tech industries such as telecommunications and biotechnology (between 0.5 per cent and 0.2 per cent compared with 4 per cent or much higher). Private sector R&D remains strong in some of the areas - such as oil and gas exploration - from which many governments have withdrawn R&D funding. RD&D spending by electric utilities appears to be declining. There is evidence from SENSER (Synergies between European and National Strategies for Energy R&D) Project; and other sources, that decreases in public funding for some longer-term technology options, such as renewable energy technologies, have caused complementary decreases in private sector R&D. The trend towards privatisation in the industry is causing R&D budgets to fall in the privatising companies.

(v) The focus of private sector R&D is shifting away from longer-term efforts to the nearer-term efforts. This shorter-term orientation reflects an increasing focus on incremental improvements to current technologies. As utilities shift their focus from 'producing power' to 'providing services', they are decreasing R&D on advanced power generation technologies in favour of information-related technologies that will help differentiate one electricity provider's product from another. They are increasingly taking on the role of 'technology users' and 'service providers' and moving away from technology development.

(vi) The trend toward nearer-term R&D is also affecting some public sector budgets. Governments have traditionally provided most of the funding for long-term R&D. As governments cooperate more closely with industry, some have shifted their R&D investments toward nearer-term efforts.

The government has an important role to play in ensuring that needed mid and longer term R&D on advanced energy technologies takes place. Governments may choose to fund such R&D directly or to put in place mechanisms and incentives to stimulate the private sector to conduct the research.

Mechanism to Support RD&D

A number of mechanisms are mentioned below:

— Formation of consortia to share the costs, risks, and benefits of RD&D. Various bilateral and multilateral arrangements are possible, involving utilities, equipment manufacturers, energy producers, and governments. Partnerships involving manufacturers can encourage the development of efficient end-use technologies.

— Joint government-private efforts can promote longer-term RD&D efforts and respond to public interest in environmental and energy security goals.

— Grants and tax incentives can play important role.

— Government policies that provide for large scale purchases or production incentives for new technologies can accelerate the introduction of new technologies.

— Regulations and Standards will encourage the introduction of technology with better environmental performance.

— Surcharge on electricity sales have been proposed as a mechanism for funding R&D consortia.

— "Green Pricing" systems allow consumers to pay a premium for their power on condition that the additional revenues be used to fund the development or purchase of 'green' technology options, chiefly renewables and efficiency measures.

— Venture Capital Funds provide a means to promote investment in advanced electro-technologies, renewable power sources, and environmental protection.

Issues having Influence on Development of Energy Sector

India's energy sector has to work in unison with the global energy sector development. Following issues will have a bearing on actions relating to technology development in India:

(i) Information technology is now taking over the traditional practices and major companies with an edge in the information technology will have the control of global markets.

(ii) Global availability of conventional energy sources will not be a constraint but the supply lines may lengthen in certain cases. South Asia is not richly endowed and as such conventional energy sources will be a constraint in the long term. This will call for effective and optimum use of locally available energy resources.

(iii) Over 1 billion poor people, without access to electricity, in South Asia will be the major priority sector for reaching electricity through renewables. Packages of practices and technology will have to be developed at a fast pace. Regional cooperation will have to be considered.

(iv) Mega Power Projects, such as coal pit head power plants, are attractive for base load. These can also have other fuel linkages - gas or liquid fuel. But, small stand-alone systems based on locally available energy resources will be crucial to reach energy in the desired form to the user. These systems have two basic advantages :

 (a) Systems will be owned and operated by the user communities or private operators in close contact with the users, and therefore able to provide flexibility of operation.

 (b) Cost of systems will be shared by the user community or private operators.

(v) Though globally sufficient funds are available for energy project development needs, availability of institutional/foreign/multi-lateral/bilateral funding for projects in South Asia is limited due to fragile economic situation. This necessitates local efforts utilizing domestic capital markets.

(vi) Transport or transmission of energy from one place to another at an economic cost has to be effected. Regional Grid for South Asia, will therefore, take precedence over other activities.

(vii) Allocations for site specific research, development and demonstration activities will have to be enhanced. This research will also focus on environmental issues. Such actions and activities alone will give a long term sustainability to energy systems.

(viii) Energy is now dealt as a 'service' rather than a commodity. It will have to be supplied in the form and quality acceptable to the consumer. The future will be much more demanding on the quality of energy supply.

(ix) International cooperation and collaboration, across the board, can improve effectiveness of performance of such systems.

The Challenge of Energy R&D

Improvement in energy technology in India will face the challenge mainly for actions relating to the following:

- lower the monetary costs of supplying energy;
- lower its effective costs still further by increasing the efficiency of its end uses;
- increase the productivity of manufacturing and agriculture sectors;
- develop high-technology energy-supply and energy-end-use products and know-how so as to reduce India's dependence on imports;
- develop renewable energy technologies to reduce cost of production, conversion and utilisation;
- develop technologies for small decentralised power generation systems;
- diversify the domestic fuel-supply and electricity-supply portfolios to build resilience against the shocks and surprises that an uncertain future is likely to deliver;
- reduce the emissions of air pollutants hazardous to human health

and to ecosystems;
- improve the safety of nuclear energy operations;
- develop technologies to slow the build-up of heat-trapping gases in the global atmosphere;
- develop packages for technology transfer to other countries in the South Asia Region;
- motivate private industry to invest in technology development and demonstration; and
- promote supply of uninterrupted quality power.

Conclusion

Greenhouse gas emission, a major environmental issue, is also giving a new direction to the transport and movement of the people. Greater dependence on individual information based systems is likely to result in lowering of travel time but will then require decentralised, reliable and efficient small systems and power packs. This may be in terms of uninterrupted power supply for computer system or for local domestic lighting or any other application.

Energy supply is now not considered a commodity but a service. In view of that the quality of energy supply and reliability become much more a proposition for the user's satisfaction rather than a simple one-time commodity sale. New markets are now emerging and these applications call for greater industry - R&D institution collaboration for basic and effective R&D leading to quality and reliable products.

REFERENCES

Conclusion & Recommendations, International Conference on Financing of the Energy Sector, Indian Member Committee of WEC, New Delhi, July 1998.

IEA Report, 17th Congress of WEC, Houston, 1998.

Inciarte, Gustavo, Societal Requirements for Increased Utilization of Conventional Resources (Technical Issues) - Review of Session 2.3, 17th Congress of WEC, Houston, September 1998.

Nakicenovic, Nebojsa, *World Energy Perspectives, WEC/IIAS A Study,* Cambridge Press, 1998.

Report of the US Federal Energy Research & Development for the Challenges of the Twenty First Century, 1997 (Extract circulated at 17th Congress, of WEC).

3

ENERGY CONSERVATION IN INDIA

Introduction

Energy conservation has received attention in India since the mid 70s. The impact of energy conservation efforts is felt at a very low pace as the commercial energy consumption per capita is low in the country, and efficient end use devices are costly. Recent rapid increase in energy demand, mainly in the industrial and the service sectors has created a renewed awareness about the economic advantage of energy conservation.

Structural changes in the economy has led to expansion of the industrial base in the country, and subsequently the increase in demand for energy. Electricity generation sector has not expanded at a desirable level. This has also supported a renewed effort on energy conservation.

The energy conservation efforts in India have to be viewed in terms of coal and lignite being the long term sustainable local energy resources. Small resources of petroleum and natural gas will be exhausted shortly and in the medium and long term the import of oil will increase.

Energy Policy

India follows an energy policy which is divided into short term, medium term and long term measures. These can be summarised as follows:

Short-Term

— maximise returns from the assets already created in the energy sector.

Paper presented at the 17th Congress of the World Energy Council on 'Energy and Technology: Sustaining World Development into the Next Millennium', at Houston, during September 13-18, 1999.

— initiate measures for reducing technical losses in production, transportation and end-use of all forms of energy.
— initiate action to reduce the energy intensity of the different consuming sectors of the economy and promote conservation and demand management through appropriate organisational and fiscal measures.
— initiate steps for meeting fully the basic energy needs of the rural and urban households, so as to reduce the existing inequities in energy use.
— maximise satisfaction of demand for energy from indigenous resources.

Medium-Term

— initiate steps towards progressive substitution of petroleum products by coal, lignite, natural gas and electricity so as to restrict the quantum of oil imports to the 1991 level.
— initiate action for accelerated development of all renewable energy resources, especially the available hydro-electric potential.
— promote programmes to achieve self-reliance in the energy sector.
— initiate appropriate organisational changes in the case of different energy sub-sectors consistent with the overall energy strategy.

Long-Term

— promote an energy supply system based largely on renewable sources of energy.
— promote technologies of production, transportation and end-use of energy that are environmentally benign and cost effective.

These policy elements were built on the hypothesis that foreign and private investments would be forthcoming at desired levels. As lack of procedures to fulfill accountability norms have resulted in poor response from investors, the performance is not in conformity with the laid down policy.

Primary Energy Endowments

Present proven recoverable reserves of different sources of commercial energy in India (1993-94) are as follows:

—	Coal	68,047	mt
—	Crude oil	756	mt
—	Natural gas	686	bcum

Primary Energy Production and Consumption Pattern

Among the indigenously produced primary commercial fuels, the relative share of oil and gas has increased from 1.2 per cent in 1950-51 to 32 per cent in 1990-91; whereas the share of coal has declined from 98 per cent in 1950-51 to 63 per cent in 1990-91. The share of coal was reflected to be 67 per cent in indigenously produced primary commercial fuels in 1995-96, mainly due to lower crude oil production.

The average rate of growth of commercial energy consumption during 1953-54 to 1990-91 varied between 4.21 per cent and 5.47 per cent. In view of shortages and restrictions, the past trend of commercial energy consumption does not really represent the growth of demand for such energy but merely reflects the growth of its actual availability.

Future Energy Demand

Expert consultations at the Asia Energy Vision 2020, organised under the auspices of the World Energy Council, agreed on energy demand projection in India, up to 2020, as given in Table 3.1.

Efficiency improvement is likely to make a substantial difference. Efficiency in generation and improvement in PLF will mean higher energy supply for the same primary input; and efficiency in use will reduce the demand of final energy. The combined effect in these 'supply side and demand side' through proper management can be close to 1.5 to 2 per cent per year or about 25 to 30 per cent by the year 2020. The ultimate demand for installed power generation capacity can come down to the range of 275,000 MW to 300,000 MW; requiring additional capacities at an average rate of 10,000 MW per year up to year 2020.

Projections upto year 2020 have been made as the present trend of primary energy dominance will be replaced by a shift towards 'Bio-Nuc' era. Nuclear energy is expected to have installed capacity upto 20,000 MW, and biomass consumption in terms of fuelwood may be about 200

TABLE 3.1 : Future Demand for Commercial Energy in India

Source	Unit	1991-92	1996-97	2009-10	2020-21*
1. Electricity	TWH	231	336	725	1300 (1745)**
2. Coal	mt	229	311	690@	1345
3. Petroleum Products	mt	57	81.2	165	335
4. Natural Gas	bcum	18.6	30.2	65	130

Source: Planning Commission, Government of India - September 1995 & September 1996.

* Projections to 2020-21 made on the basis of 'business as usual'. J.N. Maggo - R&D Needs for Energy Sector - September, 1996

** Electricity generation capacity of 1745 TWH will be required to take care of T&D losses and Plant

Load Factor. Corresponding installed generation capacity will be 378,000 MW at present day PLF and T&D loss level (1996).

@ R. K. Sachdev - International Conference on Coal - June 1996.

mt per year. However, other modern conversion technologies will be suitably adopted for biomass. Power generation from: (i) fuelwood incineration, (ii) agro-residue briquette gasification, and (iii) cogeneration in sugar mills will have proven technologies. Most conservative estimate indicates that power generation from bio fuels will have an established 20,000 MW capacity, that too is dispersed fashion. 'Bio-Nuc' era will have thrust on nuclear for base loads and bio-fuels for dispersed loads. From 2020 onwards, additional capacities in power generation may be more from 'Bio-Nuc' combination than coal based power plants. However, in terms of percentage, Bio-Nuc may contribute to 12-14 per cent of power generation.

The nuclear power scientists and engineers have done well to utilize their lean period for 'rest, relief and training', a terminology used by the army for peace time operations. Their capacity to negotiate a cost effective deal with the Soviet Union for supply of two power plants of 1000 MW each; Nuclear Power Corporation promoting joint venture operations with power producing companies; and replacement of components of the Rajasthan, Atomic Power Plant at Rawatbhata, are some of the examples as to how the Indian experts have learnt from their experience. The process of nuclear plant safety, waste disposal and nuclear radiation will need solutions, but these will evolve over time.

Energy Conservation Efforts during Eighth Five Year Plan Period (1992-97)

A National Energy Efficiency Programme (NEEP) was envisaged in the Eighth Plan with a target saving of 5000 MW of additional installed capacity in the electricity sector and another saving of 6 million tonnes in the petroleum products. The action plan prepared to achieve the target savings of 5000 MW envisaged a target of 2250 MW on the supply side and 2750 MW on the demand side.

Supply Side Savings

So far as the supply side is concerned, details of saving in additional thermal capacity in MW on account of improvement in Plant Load Factor due to R&M, reduction in thermal auxiliary consumption, reduction in T&D losses and savings in petroleum products are as given in Table 3.2.

TABLE 3.2 : Supply Side Energy Conservation Targets and Achievements

		Target (MW) (1992-97)	Achievement (MW) (1993-95)
(a)	Through efficiency improvement of generating stations with capacity utilization and reduction in auxiliary consumption.	1350	2135
(b)	Through reduction in Transmission and Distribution losses	900	1150
	Total	2250	3285

The target for oil consumption in the power sector was set at 0.3 million tonnes. Against this, the actual achievement till 1995-96 was 1.4 million tonnes.

Demand Side Savings

On the demand side, savings of 2750 MW was targeted. Though action was initiated in respect of several components on demand side, it is difficult to quantify the achievements against the target as the users of electrical equipment are numerous. However, specific energy consumption, defined as energy consumed per unit of output in the different end-use sectors has shown a declining trend as in Table 3.3:

TABLE 3.3 : Demand Side Energy Consumption Achievements

Cement	122 Kwh/tonne (1991-92)	114 Kwh/tonne (1993-94)
	930 K.Cal/tonne	880 K.Cal/tonne
Paper	1.5 Mwh/ton (1990)	1.4 Mwh/tonne (1992)
Steel (SAIL)	11.27 G.Cal/tonne (1986-87)	9.83 G.Cal/Tonne (1993-94)
Agri pump set rectification		147 Mkwh annual savings
Energy efficient lighting		About 100 MW avoided capacity

Initiatives taken by the Government during 1992-97

Energy Conservation Award

The Government of India has deciared 14th December (every year) as National Energy Conservation Day. The National Energy Conservation Award was introduced by the Ministry of Power, for the first time in 1991, for industrial units who have taken exceptional initiatives on energy conservation. Since then, such industrial units have been given national recognition through these awards. Twenty-one companies, who were the winners of Energy Conservation Awards in 1995 have indicated substantial saving as a result of energy conservation measures adopted by them. The information in respect of investment and savings by the Award Winning Companies, both in monetary and energy terms over the 3 years period has been compiled and is given at Enclosure-1 (at the end of the Chapter).

Renovation and Modernisation of Thermal Power Plants

In order to improve the performance of the existing Thermal Power Stations, a massive Renovation and Modernisation (R&M) Programme (Phase-I) was launched by the Government of India in September, 1984 for completion during the Seventh Plan period (1984-89). In view of the encouraging results achieved from the implementation of Phase-I of R&M programme of selected old thermal units during the Seventh Plan, the Phase-II for R&M of thermal power stations was taken up in the year 1990-91 for implementation during the Eighth Plan. Under the programme, 44 thermal power stations, consisting of 198 thermal units aggregating to a total capacity of 20,869 MW are covered. After the completion of the programme the benefits expected are: (i) additional generation of 7,864 Mkwh/Year, (ii) increase of 100 MW peaking capacity, (iii) life extension of 24 units of thermal units by 15-20 years. Also 75 MW of peaking capacity has been successfully re-captured after completion of works on some of the units.

Rectification of Agricultural Pumpsets

In the agriculture sector, India has about 12 million electrical pumpsets consuming 28 per cent of the electric power. A number of reports have indicated that savings of upto 30 per cent can be realised by making certain improvements and replacements in the existing pumpsets. In order to promote energy conservation, comprehensive guidelines for rectification of agricultural pumpsets were issued by the Ministry of Power in November, 1994. (Ministry of Power began sponsoring rectification of agricultural pumpsets since 1987). As per these guidelines, the Ministry of Power would normally finance the Project to the extent of 50 per cent of the total cost which would include material cost, monitoring/evaluation cost and other incidental labour charges. It is estimated that these rectifications have resulted in a saving of 417 million kwh annually equivalent to 22 MW of installed capacity.

Compulsory Energy Audit in Industry

Energy audit reports have identified the savings possible in the industrial sector. With a view to providing thrust for energy conservation and thereby improving the energy situation by optimum utilisation of power available, the Government of Kerala had, in May,1992 made energy audit compulsory for all Extra High Tension (EHT)/High Tension (HT) industrial consumers. Invoking the power given to the state governments under Section 22-B of the Indian Electricity Act, 1910, the Governments of West Bengal, Delhi and Tamil Nadu have issued similar orders. West Bengal made it mandatory for every consumer whose maximum load exceeds 500 KVA to get an energy audit conducted in his establishment once every year and to submit the report to the Power Department of the State Government. In view of this, Ministry of Power has written in November, 1996 to other state governments /UTs to issue similar orders under Indian Electricity Act, 1910, at the earliest, under intimation to Ministry of Power making energy audit mandatory for all those having a connected load of 500 KVA and above.

Replacement of GLS Bulbs by CLF Bulbs

Almost two-thirds of the lighting points in India are fitted with incandescent lamps (GLS bulbs). The fluorescent lamps of 40W/36W rating and 20W/18W rating (commonly known as tube lights) have gained considerable consumer acceptance and are reasonably priced. Also, the Ministry of Power recommended use of 36 W slim fluorescent tubes in place of 40W tubes to the State Electricity Boards to get a benefit of about 10 per cent saving in energy without compromising on the light output. The Government of India recommended to all state governments to use CLF bulbs.

A few pilot schemes on system improvement through use of LT capacitors, amorphous core transformers and remote controlled load management etc have also been taken up.

Energy Conservation Awareness Creation

On the demand side (i.e. end use side), improvements in the efficiency of energy use in various sectors are being brought about through measures such as awareness promotion, information dissemination, training, energy audits and other studies/surveys to identify areas of wastage and assess the scope for further improvement in the use of energy. In addition, subsidy for energy audits and demonstration/pilot projects on energy savings, fiscal incentives for selected energy saving devices, research and development are also under consideration.

As a result of the measures carried out by various thermal power stations in the country, there has been a continuous declining trend towards reduction of secondary fuel oil consumption in thermal power stations. On country level, the consumption of secondary fuel oil by thermal power stations is estimated by CEA to have come down from 6.77 ml/kwh in the year 1991-92 to 4.53 ml/ kwh in 1995-96, resulting in an overall saving of the order of 5.8 lakh KL of fuel oil in comparison to the 1991-92 level.

Targets for the Ninth Five Year Plan (1997-2002)

While considering the size of the country and the number of institutions/ organisations working in the area of energy conservation and efficiency and the number of end-users, it is not possible to precisely monitor the targets for the different sectors of the economy in absolute numbers. However, an attempt has been made to set some targets primarily for the supply side.

Supply Side

Central Electricity Authority, CEA has already launched Renovation and Modernisation Programme - Phase II and it is expected to result in a capacity addition of 1700 MW. CEA has also indicated a target reduction of T&D loss by 1 per cent every year during the Ninth Plan.

Demand Side

The industrial sector accounts for about 40 per cent of the total commercial energy utilised in the country; and 60 per cent of the energy in industrial sector is consumed by 6 energy intensive industries namely aluminium, fertilizer, iron & steel, cement, pulp & paper and chlor-alkali. It is important to set up norms for each industry. At the same time, it is very difficult to set up norms due to the age of the units, technology used, prevalent housekeeping measures in the units and management approach to energy conservation. However, it is proposed that there will be indicative norms for these industries for the Ninth Plan.

The norms for energy intensive industries will be set in consultation with the industry. It would be necessary to link this up with incentives and penalties that would be designed to achieve the set norms. Nodal state level agencies will be identified and will be empowered to implement energy conservation programme.

National Action Plan for Power

The Common Minimum National Action Plan for Power envisages the following in respect of energy conservation:

(a) Compulsory metering both at sub-stations and on all major feeders would be introduced. Compulsory metering of all new electricity connections as also of connections to agriculture sector exceeding 10 HP will be undertaken and completed in two years. All electric supplies would be metered by 2002A.D.

(b) Compulsory annual energy audit of large consumers i.e. 1000 KVA and above would be undertaken.

(c) Time of the Day (TOD) metering would be introduced for big power consumers for better load management.

Energy Conservation Bill

With a view to acquiring the necessary legal authority to enforce a variety of energy conservation measures, it has been decided to enact a legislation to provide a greater thrust to energy conservation activities in the country. The bill provides for setting up Central Energy Conservation Authority to facilitate and enforce efficient use of energy and this would be enabling in nature. Laying down energy efficiency standards, making energy audit mandatory, banning manufacture of the equipment not conforming to energy efficiency standards and imposition of penalty for non compliance upon the designated consumers are some of the important functions of the authority. It is proposed to table the Bill in the Parliament soon.

ENCLOSURE-1

Energy Saving by Energy Conservation Award Winning Companies in 1995 (for the period 1992 - 95)

Name	Investments (Rs. In lakhs)	Savings Electricity (lakh units)	Energy Saved/Year Fuel Oil (KL)	Gas (lakh cum)	Coal (MT)
Aluminium					
Indian Aluminium Co. Ltd Belgam	6.00	330.00	30.00	5093	—
Hindalco Industries Ltd. Renukoot	156.00	346.00	324.00	122	2327
Chemicals					
Shriram Foods & Fertilizer Industries, New Delhi.	400.00	520.00	N.A.	N.A.	N.A.
ATIC Industries Ltd., Atul	7522.00	N.A.	Specific electrical energy consumption brought down from 9489 kwh/tonne to 9419 kwh/tonne and thermal energy consumption has been lowered from 74.24HKCal/tonne to 64.68 HKCal/tonne (13 per cent).		
Chlor-Alkali					
Chemfab Alkalis Ltd Pondicherry	275.00	5.2	N.A.	N.A.	N.A.
Fertilizer					
National Fertilizer Ltd Vijaipur	597.00	90.25	—	—	34.01
Krishak Bharati Cooperative Ltd, Hazira	1250.00	295.00	Specific energy consumption reduced by 0.2575 MMKCal/MT of Ammonia (3 per cent reduction)		
India Farmers Fertilizer Co-operative Ltd, Aoula	46.85	145.78	Saving of 49704 Gcal of thermal energy (lakh/annum) per annum.		
Iron & Steel					
The Tata Iron & Steel Co.Ltd, Jamshedpur	227600.00	N.A.	Specific energy consumption reduced by 8 per cent over the last ten years. Petro-fuels reduced from 44 kg/tonne to 30 kg/tonne (13 per cent reduction).		

Rashtriya Ispat Nigam Ltd., Vishakhapatnam	620.00	N.A	Specific energy consumption has come down from 10.14 Gcals/ tonne of liquid steel in 1992-93 to 7.80 Gcal/tonne of liquid steel in 1994-95 (23 per cent) Reduction in furnace oil consumption from 11631 kl in 1993-94 to 6252 kl in 19943-95 (46 per cent reduction)		
Steel Authority of India, Bokaro	8950.00	10450.00	Specific energy consumption (including thermal and electric) has been brought down to 8.71 Gcal/tonne in 1994-95 from Gcal/tonne in 1991-92.		
Refineries					
Indian Oil Corpn Ltd, Haldia Refinery	1368.00	355.00 per annum	—	13200.00 per annum	—
Cochin Refineries Ltd Amoalamugal	175.00	127.00	—	7810.00	—
Petrochemicals					
Reliance Industries Ltd PTA Complex, Patalganga	9000.00	5408.00	Specific electrical energy consumption has been reduced from 648 kwh/tonne of Pta to 625 kwh/tonne (3.5 per cent reduction) while the specific consumption of thermal energy has been lowered from 2.53 MKCal/tonne to 1.95 MKCal/ tonne (23 per cent reduction)		
India Polyfibres Ltd Lucknow	12.67	53.09	Energy consumption as percentage of manufacturing cost from 5.89 per cent in 1992-93 to 4.86 per cent in 1994-95.		
Pulp & Paper					
Seshasayee Paper and Boards Ltd, Erode	N.A.	Achieved monetary saving of 5 per cent over three year period	N.A.	N.A	N.A.
Ballapur Industries Ltd Yamuna Nagar	N.A.	N.A.	Specific electrical energy consumption reduced from 1186 kwh/tonne in 1991-92 to 1092 kwh/tonne in 1994-95 (8 per cent reduction) and thermal energy consumption from 4.606 MKCal/ tonne to 4.472 MKCal/tonne		

			during the same period (3 per cent reduction).		
Textiles					
Kesoram Rayon, Hayasaral (Hooghly)	99.00	89.00	39.00	—	810.00
J.K.Synthetics Ltd	16.00	190.00	Specific electrical energy consumption reduced from 3929 kwh/tonne in 1992-93 to 3431 kwh/tonne in 1994-95 (13 per cent reduction) and specific thermal energy consumption reduced from 1.261 MKCal/ tonne to 0.973 MKCal/ tonne in the same period (23 per cent reduction).		
The Deccan Cooperative Spinning Mills Ltd, Ichalkaruji (Kolapur)	7.00	35.00	15.00	—	—

REFERENCES

Chaturvedi, Pradeep, Energy Conservation Perspectives for India, National Workshop on Energy, IE(I) and RGF, New Delhi, September 8-9, 1995.

Energy Conservation Award Book 1999, Energy Management Centre, 1996.

Eighth Five Year Plan (1992-97), Planning Commission, Government of India.

Maggo, J.N., *R&D Needs for Energy Sector,* Department of Science & Technology, Government of India, New Delhi, October 1996.

Mittal, U.K. Energy Conservation Measures for SAIL, National Workshop on Energy, IE(I) and RGF, New Delhi, September, 8-9-1999.

Raghupathy, S., Energy Conservation: The catalytic Role of Industrial Associations, National Workshop on Energy IE(I) and RGF, New Delhi, September, 8-9-1999.

Srivastava, R.N., Indian Electricity Supply: Issues and Options for Tomorrow, Asia Energy Vision 2020 - IE(I), New Delhi, November 14-16, 1996.

Sachdeva, R.K., Future Demand for Coal, International Conference on Coal, New Delhi, June 1996.

Tiwari, S.K. Energy Conservation; A Perspective, National Workshop on Energy—IE(I) and RGF, New Delhi, September 8-9-1999.

4

ENERGY AND ENVIRONMENT MODELLING IN INDIA

Introduction

Energy and environment issues attracted global attention in the year 1972 - energy for sudden price rise; and environment for pollution problem. These events brought in a new thinking on energy and development issues. The United Nations Conference on Human Environment, held at Stockholm in 1972, was an eye opener to the developed as well as the developing countries. Environmental degradation and depletion of natural resources were seriously considered. Poverty was projected as a major cause for environmental degradation. Poverty embraced the lack of capacity to buy services that would ensure cleaner development or can result in cleaning up the degradation effects. 1972 also witnessed the first oil shock when the prices of crude oil considerably shot up in the international oil market. The present energy and environment concerns are deeply interrelated. Environment issues have shifted the focus in the energy sector to non-conventional and nuclear resources in preference to fossil resources. However, declining oil prices and free trade has retained the supremacy of oil in the energy supply.

Economic Situation in India

The rate of growth of consumption of commercial energy has been about 5.5 per cent per annum during the last five decades (1947-1997).

Over this same period the use of non-commercial energy sources has declined from 71.6 per cent in 1953-54 to 34.1 per cent in 1996-97.

Paper presented at the IEA-India Workshop on 'Data Requirements for Energy Policy - Statistics, Modelling and Indicators of Energy Use', organised by the Central Electricity Authority on May 13-14, 1999 at New Delhi.

Even at 34.1 per cent it is providing one-third of the primary energy needs of the population. Any indication of economy energy relationship purely on the basis of the commercial sources of energy will not be a true representation of the fact.

Another important aspect that needs to be considered is that the village and cottage industries are mostly based on non-commercial sources of energy. Contribution of such industries is recognised not only for the local production, but also in some cases for exports. The contribution of rural and unorganised sector to the national economy is significant and is partly reflected in the energy intensities based on total primary energy use. The present practice does not reflect energy truly in the GDP accounting.

The elasticity of total primary energy consumption for the period 1953 to 1996 with respect to Gross Domestic Product (GDP), at constant price, works out to 0.78 per cent as compared to 1.25 per cent for commercial energy. The elasticity of total final energy consumption works out to 0.67 per cent as compared to 1.14 per cent for final commercial energy consumption for the same period.

These energy elasticity projections do not take into consideration any environmental impact due to extraction and utilization of natural resources. This impact assessment is necessary to work out the actual cost of utilisation of energy resources. Energy elasticity figures could vary if the environmental impact is also internalised.

Economy Energy Relationship

Natural Resource Accounting will be the most important tool for establishing relationship between economy and environment. This has direct relevance for the energy sector. It is interesting to note as to what is involved in natural resource accounting.

Natural Resource Accounting

In an environment-economic accounting system, increasing emphasis is to be given to the natural environment that has been or could be affected by human activities. The assets of natural environment that are directly or indirectly, actually or potentially affected by human activities are called natural assets. Natural assets consist of biological assets (produced or wild), land and water areas with their ecosystems, subsoil assets and air. To develop such an environment-economic accounting system it is necessary to maintain natural resource accounts. However, no such system has been developed so far due to lack of proper methodology.

The economic development of a country or region is generally expressed in terms of the growth of its income. The value of the final product, excluding the value of inputs used in the process of production, is termed as the Gross National Product (GNP). However, these indicators of economic development do not take into account the use and depreciation of the renewable or non-renewable natural resources. But, the growing scarcity of these resourses has forced the policy-makers to develop the natural resource accounts. Following other countries India too has given due thought to this phenomenon and initiated a pilot project on Natural Resource Accounting in Goa. After the development of a suitable methodology, this may be extended to other states as well.

Environmental Issues of Energy

The environmental problems associated with the energy sector have to be viewed separately in terms of use of commercial sources of energy and the renewable sources of energy, mainly the hydro resources and the traditional biomass. The environmental impact of such resources will affect some geographic locations more than the others. Availability of such resources is skewed, as detailed below:

Distribution of Energy Resources

Commercial Energy Resources and Regional Distribution

Coal and hydroelectric resources are the mainstay for India's primary resource support for secondary energy purposes. Oil and natural gas resources of the country are limited. Nuclear resources, on the other hand, are modest. The coal resources are of poor quality having high ash content. Besides the above, the distribution of the basic resources, i.e., coal and hydroelectric resources in the various regions, are rather skewed. The eastern region accounts for over 70 per cent of the total coal resources; the western region accounts for over 70 per cent of the hydrocarbon reserves; and the northern and north-eastern region account for over 70 per cent of the total hydro potential of the country. The southern region which has only 6 per cent of the coal reserves and 10 per cent of the hydroelectric potential, possesses most of the lignite deposits of the country.

The regional distribution of the primary commercial energy resources are given in Table 4.1 below:

TABLE 4.1: Primary Commercial Energy Resources (March 1997)

Region	Coal (BMT)	Lignite (BMT)	Crude Oil (MMT)	Natural Gas (BCM)	Hyrdo Potential at 60% LF (MW)	Power Energy (TWH)
Northern	1.1	1.1	—	4	30135	225.00
Western	48.2	0.5	584	497	5679	31.40
Southern	13.1	25.9	—	—	10763	61.80
Eastern	141.4	—	—	—	5590	42.50
North Eastern	0.9	—	148	159	31857	239.30
All India	204.7	27.5	732	660	84044	600.00

Source: Planning Commission, Government of India (1998) and ONGC (1999).

BMT = Billion Metric Tonnes. MMT = Million Metric Tonnes.

BCM = Billion Cubic Metre.

Note: Gas reserves are now identified at 717.3 BCM (at the end of 1998) distributed under control of ONGC - 525 BCM; OIL - 94.5 BCM & Joint Ventures/Private companies - 97.5 BCM.

Thus the environmental concerns due to extraction of coal will be pronounced for the eastern region, followed by the western region but of no significance to other parts. Similarly the environmental impact due to drilling for oil will cause concern in the western region and the north eastern region, but not in other regions of the country.

Hydro potential is largely located in the northern and north eastern regions, mainly in the hilly and ecologically fragile regions.

Non-conventional Energy Sources

The country is endowed with a large potential of non-conventional energy sources. The potential availability and extent of exploitation as on March 31, 1998 are given in Table 4.2 below :

Traditional Non Commercial Primary Energy Sources

Fuelwood is an important source of energy in the rural and, to some extent, in the urban households. The total forest area in the country has been assessed at nearly 75 million hectares. The growth in the demand for wood for industry, construction and for domestic use for cooking etc

TABLE 4.2: Renewable Energy Potential In India

Source/Technology	Potential Availability
Biogas Plants	12 Million
Million	
Biomass-based Power	17,000 MW
Efficient Woodstoves	120 Million
Solar Energy	5 x 10 Whr/Year
SWH	
Small Hydro	10.000 MW
Wind Energy	20,000 MW
Energy Recovery from	
Urban & Industrial Wastes	—
Ocean Thermal	50,000 MW
Sea Wave Power	20,000 MW
Tidal Power	9,000 MW

Source: Ministry of Non-Conventional Energy Sources, Government of India, Annual Report 1997-98.

SPV = Solar Photovoltaic. SWH = Solar Water Heater. SC = Solar Cooker.

has exceeded the annual incremental availability of wood from local sources and social forestry schemes. This has resulted in the gradual denudation of tree cover and consequent adverse impact on the environment. Recent efforts in energy plantation and inter-fuel substitution has arrested and reversed this trend.

Fuelwood is supplemented by dung and crop residues in meeting domestic energy needs in the rural areas. The annual availability of wet dung is estimated to be about 960 mt. Greater use of dung cake as a source of energy reduces the availability of dung as a valuable organic manure. Total availability of crop residues is estimated to be 450-500 mt. The draught animal population in the country has been estimated at 70 million and animal energy continues to be used in agricultural operations and for rural transportation.

Environmental Impact

Environmental Issues of Commercial Energy Use

Commercial energy sources create environmental problems in two ways-at the stage of extraction, public mining and later during utilisation in industries. In the States of Bihar, West Bengal and Orissa, open cast mining of coal, limestone and iron ore has contributed to the degradation of soil and corresponding run-off of water and soil. Coal suffers from quality problems arising from the high ash contents which not only contributes to lower industrial activity but also creates pollution problems as the thermal power plants in India denote. Its use is associated with carbon dioxide emissions and fly ash production.

Oil exploration being carried out in India under mainland and off-shore conditions has its peculiar characteristics. Oil spills during extraction and transportation are often the cause for sea water pollution. The increased use of petroleum in automobiles and as feed stock in industry has also added to pollution problems, primarily by way of emission of sulphur dioxide and carbon monoxide etc.

Uranium and thorium extraction has often raised alarm for radiation effect on the workers and residents of the mining area. Its conversion into appropriate neclear fuel and disposal of spent fuel has always been a matter of concern for its harmful environmental impact.

Environmental Issues of Renewable Energy Use

The problem with hydro power projects on the other hand lies in its natural resource depleting effects, particularly in sensitive bio-diversity areas. Most of the hydro potential exists in hilly and ecologically fragile areas. Impact on flora and fauna has to be taken into consideration for siting a plant, even if the emission effect of hydro power generation is negligible. This has also been a cause of concern for the environmentalists leading to extensive delays in execution of hydro power projects.

Use of traditional biomass is accompanied by gradual denudation of tree cover. However, recent studies have reflected that fuelwood consumption is not a major cause for forest denudation. Fast growing fuelwood trees have substantially supported decentralised fuelwood production meeting the fuelwood needs at site. Larger growth of such energy plantations will also improve the carbon sequestration possibilities.

The energy environment interaction has to be viewed in this background. Each and every element mentioned above will have its impact. Energy Environment Modelling, internalising all costs, will call for a number of identifiable determinants that can be fully supported by generally acceptable energy and environment data. Some efforts have been made in India to develop a firm basis for developing the models reflecting true costs. The following is a review of some of these efforts:

Environment Legislation in India

Essential elements of environment are reflected in the Constitution of the Republic of India, adopted in 1950. The Directive Principles of State Policy reflect the commitment of State to protect environment with regard to forests and wildlife. Initiation of the present day organisational framework for environment programmes in India go back to the 1970s with the establishment of the National Committee on Environmental Planning and Coordination by Smt. Indira Gandhi, the then Prime Minister, immediately following the Stockholm Conference. The Committee was gradually to evolve into a Department of Environment in 1980; and subsequently into a full-fledged Ministry of Environment and Forests.

The National Conservation Strategy and Policy Statement on Environment and Development, adopted in June 1992, forms the basis for internalisation of environmental considerations in the policies and programmes of different sectors. It also emphasizes sustainable life styles in the proper management and conservation of resources. The Policy Statement on the Abatement of Pollution, 1992, emphasises the government's commitment to prevent further deterioration of the environment.

Establishment of industrial structure and power generation units focussed attention on pollution due to air, water and land resources. These pressures are reflected in the growing incidents of air and water pollution. In fact the Factories Act, 1948 with a thrust on occupational and civic safety was supplemented by the Water (Prevention and Control of Pollution) Act 1974 and the Air (Prevention and Control of Pollution) Act, 1981. These were followed by the Environment (Protection) Act of 1986.

Air and Water (Prevention and Control of Pollution) Acts

The Water (Prevention and Control of Pollution)Act, 1974

The situation regarding quantum and quality of water in the country is quite serious. The Water (Prevention and Pollution) Act, 1974 provides for the prevention and control of water pollution and the maintenance or restoration of wholesomeness of water. As such, all human activities having a bearing on water quality are covered by this Act. The industrial pollution control has been assigned the highest priority. Trade effluent in this Act is defined as any liquid, gaseous or solid substance which is discharged from any premises used for carrying on any trade or industry.

Mining is also treated as an industry under the Water (Prevention and Control of Pollution) Cess Act, 1977. As such, the provisions of the Act (Section 25 and 26) would be applicable to the coal washeries and industries using coal such as coal fired power plants. While Section 26 relates to discharge existing before the commencement of the Act, Section 25 covers new outlets and new discharges. In all cases, consent to bring into use any new or altered outlet for the discharge of trade effluent has to be obtained from the concerned State Pollution Control Board.

This Act also has relevance for development of sites for hydro-electricity generation.

The Air (Prevention and Control of Pollution) Act, 1983

This Act covers the prevention, control and abatement of air pollution. But, the consent requirement is restricted to industrial plants operating in air pollution control area to be declared by the concerned State Governments. Section 22 of the Act stipulates that no person operating any industrial plant, in any air pollution control areas, should discharge or cause or permit to be discharged the emission of any air pollutant in excess of the standards laid down by the State Pollution Control Board under Section 17(1) (g).

The consent issued to coal utilisation processes and units may include the following conditions:

— Control equipment of approved specification to be installed.
— Alteration/replacement of control equipment.
— Erection/re-erection of chimney of prescribed specification.
— Condition regarding emission standards.
— Stipulated time frame to comply with the above.

The Environment (Protection) Act, 1986

This is an umbrella Act providing for the protection and improvement of environment and for matters connected therewith. It provides that:

"No person carrying on any industry, operation or process should discharge or emit or permit to be discharged or emitted any environmental pollutant in excess of such standards as may be prescribed." Section 15 of the Act provides for punishment—imprisonment and fine for violation of the provision of the Act.

Standards for emission of air pollutants and discharge of effluents from coal based power plants have been notified under this Act.

The Environment (Protection) Act, 1986, has been amended vide Ministry of Environment and Forests Notification (The Gazette of India: Extraordinary Part II - Sec. 3 (i) dated the 22nd April, 1993 Annexure-I).

Environmental Impact Assessment

Environmental Impact Assessment (EIA) is an important decision making tool both for resource optimisation as well as for maintaining ecological balance. Impact assessment of development projects was initiated in 1977-78 with river valley projects; and subsequently, other sectors of development like mining, thermal power plants, etc were also brought under the purview of Environmental Impact Assessment by the Union Ministry of Environment and Forests. All thermal power projects irrespective of their size or location, are presently required to obtain prior environmental clearance before the site is finalised and the project is taken up.

Mining and thermal power projects requiring use of forest land also require approval under the Forest (Conservation) Act, 1980 for diversion of forest land to non-forest purposes. Such proposals are evaluated by an Advisory Committee under the Chairmanship of the Inspector General of Forests in the Ministry of Environment and Forests.

Need for Environmental Assessment

Sustainable development is thus one aspect of the need of re-appraising the environment. The high cost in terms of protecting human health due to air, water pollution; damage to flora and fauna; recreational and aesthetic aspects with improved living standards also contribute to a greater awareness for maintaining environment quality.

The EIA includes:

(i) An assessment of the environmental effects.
(ii) Prediction of changes in environmental quality as a result of direct or indirect consequences of the activity.
(iii) An attempt, to consolidate environmental effects to a common basis with economic costs and benefits for comparison.
(iv) Comparisons of various alternatives by which desired objectives can be realised and identification of that alternative which could

optimise economic and environmental benefits as compared to cost.

(v) An implementable plan for management and monitoring of environmental aspects during and after the construction phase.

Environmental Impact Statement

Cost-benefit analysis, cause-effect and environmental impact analysis is carried out before putting up proposals for project/scheme. The main purpose is to ensure that impact of alternative action are evaluated and considered in project planning. The sequence of an Environment Impact Statement (EIS) is:

(i) State major objectives of the project.
(ii) Identify technological possibilities of achieving the objective.
(iii) Specify one or two actions for achieving the objective.
(iv) Evaluate characteristics and conditions of the existing environment.
(v) Analyse the likely impact of the actions on the existing environment including prediction of important environmental factors.
(vi) Recommend a suitable way of action.

Environmental Audit

The environmental evaluation of industrial units, including thermal power projects focuses on, amongst others, the status of process and technology as well as management practices being adopted so that the inputs in the form of raw materials, energy and manpower can be minimised and the benefit maximised. Environmental audit is a management tool for the managements to evaluate the performance of their operations and take stock of the inputs for effecting mid-course correction, if any. All the industrial production units are now required to submit environmental audit reports every year as per Notification No. 120 published in Part II Section 3 Sub-Section (i) of the Gazette of India Extraordinary dated March 13, 1992.

Environment Statistics

The environment statistics are expected to provide facts necessary for government administration and policy-makers towards its effectively implementing the policies and programmes. Environment statistics is a difficult and complex subject. A careful development of methodology and making the process functional has taken place for two decades.

Environment Statistics was discussed for the first time at the Fifth Conference of Central and State Statistical Organisation (CSSO) held at New Delhi in 1981. The subject was further discussed and a final decision was taken in 1985.

A Working Group on Environment Statistics was constituted under the Chairmanship of DG, Central Statistics Organisation in 1986 with the following terms of reference:

(i) To identify the major areas of environmental concern on which data need to be developed; and
(ii) To identify the parameters pertaining to the above areas; assigning priorities and suggesting the methodology for compilation of data on these parameters.

The Working Group submitted its report in November 1990. The report indicated that environmental problems arising in India can be attributed to conditions of poverty and underdevelopment besides due to the development process itself.

A Steering Committee on Environment Statistics was constituted in November 1996 under the chairmanship of the then DG, CSO, with the terms of reference as under:

(i) Preparation of framework for the development of Environment Statistics.
(ii) Drawing a list of Environment Statistics and core environmental indicators that can be completed from the available basic information on natural resources use/depletion, soil, water and air quality, quality of life in population centres etc.
(iii) Compilation of identified and indicators based on the already available basic information for dissemination.

The complexity of laying down the norms for collection of data on environmental statistics can be realised from the fact that the Steering Committee had to meet thrice to provide guidelines for setting up an efficient environment statistics system.

Project on Institutional Strengthening and Collection of Environment Statistics

India being one of the developing member countries of the Asian Development Bank, it was approached by the latter in 1994, to implement a project on 'Institutional strengthening of Environment Statistics in India'. The Department of Statistics was selected as the implementing agency. Following activities were envisaged under the project.

(i) To compile a Compendium on Environment Statistics.
(ii) To hold an Inter Regional Training Course on Environment Statistics.
(iii) To hold a National Level Workshop on Environment Statistics.
(iv) To develop a Framework for Environment Statistics.

A Compendium of Environment Statistics - 1997 was published by CSO. The Compendium provides a glimpse of the present scenario of environment degradation, its causes and reasons for concern. It provides the necessary base to bring out the magnitude of the problem. The compendium consists of seven chapters. The first two chapters give a general introduction to environment, its degradation through different sources and their impact on human health and the development of environment statistics in India. The remaining five chapters are on Biodiversity, Atmosphere, Land/Soil, Water and Human Settlements.

Environmental Indicators

Environmental indicators have been identified to provide a clue to a matter of larger significance that makes perceptible trend or phenomenon, which is not immediately detectable.

The desirable qualities of indicators are as follows:

(i) It is a reliable statistical measure.
(ii) It is measured in comparable fashion over time and space.
(iii) Its changes can be assessed against a scientifically valid norms or standard.
(iv) It bears upon a clearly identifiable social goal.

Ideally, the index should be statistically robust and should characterize the relationship between environmental qualities and resource depletion for which it is designed to monitor. Index should also be replicable. It should say something about how environmental quality compares to standards. The index should allow the comparison of environmental quality in geographical and ecological dissimilar places, for example, the environmental qualities of two countries may be deteriorating, but one may be destroying the ambient air quality in its major cities while the other is depleting its resource base. Both countries are involved in a different process of environment change, but both are debasing the environment. An environmental quality index should be capturing both forms of deterioration.

The United Nations Environment Programme has suggested the following criteria for selection of indicators.

(i) Primarily national in scale and scope.
(ii) Relevant to main objective of assessing progress towards sustainable development.
(iii) Understandable in that they are clear, simple and unambiguous.
(iv) Realizable within the capacities of national governments given their logistic, time, technical and other constraints.
(v) Conceptually well founded.
(vi) Limited in number; remaining open ended and adaptable to future development.
(vii) Broad in coverage of Agenda 21 and all aspects of sustainable development.
(viii) Representatives of an international consensus to the extent possible.
(ix) Dependent on data which are readily available at reasonable cost/benefit ratio, adequately documented of known quality and updated at regular intervals.

Presently, indicator functions are more closely focused on the interaction between the natural environment and socio-economic decision making. According to the World Bank, an environmental indicator is meant to be used in conjunction with socio-economic indicator and oriented towards the specified aims of the institution. They hope to use indicators in a bold effort to squeeze more policy relevant from the data, that exist with the objective of improving current practices in the analysis of traditional projects, public expenditures and investment reviews. The indicator should also set the context for economic sector work as well as sectoral and structural adjustment operations. The indicators will then be used to identify the environmental impacts of national economic policies to choose policy instructions for environmental protection and to evaluate the environmental economic impact of projects.

The central statistical organisation has identified that several frameworks have been developed to organise indicators for sustainable development and especially environmental indicators. One way to capture most of the approaches and listings of environmental and sustainable indicators is to combine the concerns of potential data users as reflected in United Nations Conference on Environment and Development's Agenda-21 with the framework for environmental data production, such as the United Nations Framework for the Development of Environment Statistics, (FDES) 1994. For the purpose of environmental indicators, Agenda 21 issues can be arranged in a number of clusters, referring to different chapters of Agenda -21. Table 4.3 gives the clusters in the first column (only those having a bearing on energy resource, supply and use). The information categories of the FDES are shown in the column headings of the table.

TABLE 4.3: Core list of environmental and related socio-economic indicators

	FDES Information Categories			
1	2	3	4	5
Agenda 21 Issues (groups of clusters)	A.Socio-economic activities events (pressure/driving force)	B.Impacts and effects (part of state)	C. Responses to impacts (response)	D. Inventories stocks, background conditions (part of state)
Economic Issues	Real GDP per capital growth rate*. Production and consumption patterns. Investment share in GDP	EDP/EVA per capita, Capital accumulation (environmentally adjusted)	Environmental protection expenditure as % of GDP. Environmental taxes and subsidies as % of government revenue	Produced capital stock
Social/Demographic Issues	Population growth rate*. Population density*. Urban/ rural migration rate*. Caloric supply per capita	% of urban population exposed to concentrations of SO_2, particulates, Ozone, CO and Pb. Infant mortality rate*, Incidence of environmentally related diseases*.		Population living in absolute poverty. Adult literacy Rate*, Combined primary and secondary school Enrollment ratio. Life expectancy at birth*. Females per 100 males in secondary school
Air/Climate	Emissions of CO_2, SO_2 and NO_2. Production of ozone destroying substances*	Ambient concentrations of CO_2, SO_2, NO_2 and O_2 in urban areas*. Air quality index	Expenditure on air pollution abatement, Reduction in consumption of substances and emissions	Weather and climate conditions*
Other Natural Resources Biological Resources	Annual roundwood production*,Fuelwood consumption per capita*	Deforestation rate* Threatened. Extinct species*.	Reforestation rate*, Protected forest area as per cent of total land area*	Forest inventory* Ecosystems inventory, Fauna and flora inventory.
Mineral (incl. Energy) Resources	Annual energy consumption capita*, Extraction of other	Depletion of mineral resources (% of proven reserves)*		Proven mineral reserves*, Proven energy reserves*

1	2	3	4	5
	mineral resources*	Lifetime of proven reserves		
Human Settlements	Rate of growth or urban population*, per cent of population in Urban areas* motor vehicles in use.	Area and population in marginal settlements*, Shelter Index, % of population with sanitary services*	Expenditure on low-cost housing*	Stock of shelter and infrastructure

* In this Table, The indicators with (*) mark are proposed to be compiled by the United Nations Statistical Division.

Determinants for Future Energy Systems

The World Energy Council in its report 'Global Energy Perspectives', published in 1998, has identified a number of determinants for future energy systems. This is a list of determinants generally applicable to various countries. Certain issues relevant to India are not appropriately covered by these elements. Poverty alleviation is significant to India and reflects the most urgent socio-economic need of the country. Also, a reflection of energy elasticities or final form of energy will truly reflect the contribution of renewable sources of energy.

Thus following determinants will have to be considered for developing indicators in the energy environment model for India.

(i) Population growth
(ii) Economic growth
(iii) Energy intensity improvement
(iv) Technological change
(v) Resource base
(vi) Environment
(vii) Poverty alleviation
(viii) Final form of energy

With a view to present the data in a universally acceptable format it will be recommended to follow universally accepted determinants.

Population Growth

In different scenarios for projection of energy supply and consumption, the WEC has used a medium projection for the future population. That same pattern of growth and population was used in all cases, so that the differences that emerge among different cases are easily linked to differences in their energy systems and their driving forces. The UN and the World Bank projections have certain variations. IIASA, in 1996, revised its model to include subjective probabilities to possible future trends in fertility, mortality and immigration in different regions of the world. This is an important requirement for developing a model for India. It has been observed that if population variations are not independent of energy development, then the addition of population variants overlays the energy pictures with demographic debates.

A control on population growth will indicate faster economic growth and better and efficient energy utilisation and lower energy consumption. The rural to urban migration will be a major factor in deciding on the energy consumption pattern. Providing adequate and clean energy services to a population living predominantly in urban areas will be a challenging task.

Economic Growth

Economic development and growth are fundamental pre-requisites for achieving an increase in living standards with vigorous population growth in the country. The sustained high growth rate will ensure an overall improvement in technological and management practices and systems operation. This will all add up to more efficient energy consumption pattern, though requiring higher levels of absolute energy consumption. In addition to the traditional 'tangible' factors driving economic development, such as technology, capital, natural resources and trade possibilities; 'intangible' factors such as the quality of human capital and institutions, limited income disparities and social cohesion increasingly explain the historically uneven paths of economic developments. The time and speed of economic restructuring will be dependent on political stability, good governance and appropriate linkage of industry and trade with government policies. Additional components can be defined as promotion of free exchange of goods and technology, favourable investment climate and high level of education and social protection for the work force. The energy sector will have to compete the much needed investment funds in a period of increasing competition, where substantial funds are also needed for the health, communication and transport infrastructure needed in growing economy.

Energy Intensity Improvement

Energy intensity improvements have a significant impact on final energy availability from the same infrastructure without any proportional investment. The causes for such an approach are complex. They include the technological improvements in individual energy end use and conversion components. For example, a more efficient power plant. They also include structural shifts in energy systems, such as a shift from coal fired electricity generation to gas fired combined cycle plants. They also include the fuel substitution at the level of energy end use, like the replacement of fuelwood with electricity or LPG. They include economic shifts from more or less energy intensive activities They include changing pattern of energy end use and, ultimately, changing life styles.

Not every change in every one of these categories represents a decrease in energy intensity. But taken together, the overall trends are persistent and pervasive. The economic developments always outpace the increase in energy use, leading to substantial reduction in energy intensities. The energy intensities are measured both in terms of total energy divided by GDP and as commercial energy divided by GDP. Commercial energy intensities increase during early phase of industrialisation as traditional and less efficient energy forms are replaced by commercial energy. When this process is ccmpleted, the commercial energy intensity initially peaks and then declines. The energy intensity measurement in terms of the total energy consumption will be more appropriate for conditions in India as 34 per cent of the energy supply is from non-commercial sources of energy, that normally does not figure in energy planning.

Technological Change

Technology has been at the heart of economic development and change. Technology is considered crucial for improving performance and reducing cost. New environment friendly technologies have moved ahead with great promise and potential in the energy sector. Long term impact of technology have to be considered. Following five characteristics of the process are identified.

(i) Innovation and experimentation with new technology in expectation of economic opportunities and profits.
(ii) Innovation and experimentation as a hedging strategy against uncertainty of future development.

(iii) Continuous change and improvement through research, development and demonstration and actual hands on experience.
(iv) Diffusion and substitution that is the gradual adoption and spread of new technologies in time and space and their interaction with existing technologies.
(v) Impact (social, economic resource and environmental) and their feedback on technological change.

Technology cost and performance improve with experience, and there is a pattern of such improvement common to most technologies.

Resource Base

Energy resources have obvious implications for future energy system development. Identified occurrences have the highest geological assurance, followed by inferred and speculative occurrences. Reserves are those occurrences that are known and are recoverable with present technologies at prevailing market conditions. Resources are occurrences that are distinct from reserve in that they have less certain geological assurance or lack present economic feasibility, or both. Changing marketing conditions, innovations and advances in geo-sciences can transform resources into reserves. The resource base however, depends on the knowledge of technology and the economic cost at which they can be explored. Nuclear energy's future will depend on, among other things, how current controversies concerning safety, waste disposal and proliferation are resolved. The renewable energy resources are dispersed and diffused though they have such advantages like the environmental advantage but they cannot provide energy services on a continuous long term basis. Thus, resource availability has a direct relationship to the level of environment concerns and the pricing of energy.

Environment

The type and extent of pollution are closely related to the degree of economic development and industrialisation. To some extent, economic development enables societies to successfully address environmental problems of both poverty and industrialisation. But economic development and affluence can also generate new environmental problems such as waste disposal and possible global warming. Local impacts have been and continue to be a major environmental concern, related to energy use. Despite major reductions in local air pollution, the developed world is more concerned about problems of long term global issues including the climate change. The situation with transport related air pollution is dramatic in many cities. Industrialisation and thermal power generation have added the carbon dioxide and sulphur dioxide emissions on a large scale. The possibility of controlling emissions vis-à-vis efficiency improvement has an important element of concern.

Poverty Alleviation

Poverty alleviation is an important determinant in the Indian energy scene. Approach for poverty alleviation is totally different from that of building infrastructure for economic development. The measures aimed at poverty alleviation will include low cost, low level technologies which may or may not be of the highest efficiency level but will be a good employment provider. The role of non-commercial sources of energy will be better than that of the commercial sources of energy for one simple reason, that the non-commercial sources are widely dispersed and available nearest to the needy. Such an approach will not have the best efficiency in management or application, but it is urgently required for almost 40 per cent of the population in India. Inspite of consistent efforts of the Rural Electrification Corporation to reach out electricity to the poorest, not even 30 per cent of the rural population is really benefiting from this facility as they do not have the minimum income to access the electricity supply available to them.

Final form of Energy

All the determinants and the indicators have been developed on the basis of the primary form of energy and have therefore, left out any consideration for providing inputs that may include the contribution of bio fuels and other renewable sources of energy in the overall energy supply. Even the energy policy studies have often ignored the role of non-commercial sources of energy. These sources contributed to 34 per cent of energy supply in the country. As the globe is now moving towards focusing attention on the final form of energy for taking various policy decisions, serious consideration has been floated to convert solid biomass fuel into liquid, gas and electricity to provide the user his preferred form of energy. In such a case the input-output matrix will have to develop on the basis of the intensities and contributions of the final form of energy. Some effort has been made and it will require a bigger effort to follow.

Energy Environment Model-Efforts in India

No effective or foolproof energy environment model has been developed in the country. Certain efforts at energy modelling were tried in the following institutions.

(i) Planning Commission, Government of India.
(ii) Indira Gandhi Institute of Development Research.
(iii) Tata Energy Research Institute.
(iv) Indian Institute of Science, Bangalore.
(v) Indian Institute of Management, Ahmedabad.

These exercises focused mainly on modelling for energy demand and supply and consumption patterns. Details of the exercise carried out at the national level by the Planning Commission are given below.

Planning Commission

The Planning Commission, being the agency responsible for preparing both the five year and annual plans for the various sector of the economy is the agency to reflect the short term and the long term relationship for energy supply and consumption pattern and energy environment relationship. The sectoral estimates prepared by the expert working group are compiled by the Perspective Planning Division. The sectoral demand estimates are expected to be consistent with the growth profile of the economy. There are models on energy demand and supply optimization that have been developed by the Planning Commission and other agencies.

A study was carried out during the period 1986 to 1989 in the Planning Commission by Dr. R.P. Sengupta. An integrated system of model was developed in the study for comprehensive quantitative analysis of the issues related to Perspective Planning Policy for the commercial energy sector of India. The report observed, 'With the beginning of the era of liberalisation of the Indian economy in the mid 80s the consistency exercises of macro-economic, multi-sectoral projections were no longer considered adequate to meet the requirements of planning. Sectoral planning was conceived to achieve the objective of improvement in investment productivity by optimising choices in the relevant areas.

The merit of any of these models will depend on the extent to which it captures the problem of ground reality in the concerned relationship. The total model system cover the whole gamut of issues and choices relating to the exploration and exploitation of the primary energy resources, the transportation of these resources to the location of conversion, the conversion of the resources into energy products, the transportation of the energy products to the consumption points and finally the pattern of consumption. The demand projection model in the system assume alternative macro-economic targets of growth, income distribution and conservation of energy and environmental resources. It also worked out the impact of these factors on the energy demand in the terminal years through the energy-economy inter-relation'.

This Report focussed on the commercial energy sources and did not consider the contribution of non-commercial energy sources. An updated and revised report, released in 1993 identified the need for further activities to establish energy-environment relationship and to internalise environmental costs.

Model for Reference

The World Energy Council uses the IIASA Model for short term and long term energy modelling. To achieve its objective, IIASA uses a number of models that are then combined into one single overall model. The Regional Air Pollution Impacts Model, RAINS, and Model for the Assessment of GHG Induced Climate Change, MAGICC, will be relevant and can be suitably adopted for energy environment modelling framework with reference to India. IIASA has carefully included the scenario definition and evaluation to include economic development, demographic projections, technological change, international prices, environmental policies and energy intensity. Other elements relevant to India can also be suitably re-worked into existing models.

The database referred includes energy, economy, resources and technology inventory. Efforts at data identification by the Central Statistical Organisation is an important exercise and can provide a number of information relevant for energy environment model development in the country.

Suggestions on Crucial Issues to Ensure Energy Demand Supply and Environmental Improvements

The following recommendations have been identified by experts for consideration:

(i) *Need for improved energy data:* The energy data is an

inadequate historical time series (in particular, it is too short to be used for longer term projections and the sectoral breakdowns are too broad). Also, there are difficulties in using them for international comparisons with the energy statistics published by other organisations. Sectoral breakdowns are necessary for relevant energy analysis.

(ii) *Effective energy pricing practices:* Specific reference to prices is difficult as appropriate price variables from published data is not available. Very often the energy price variables differ in different publications of the same organisation. Also energy prices do not reflect the market trends because of subsidies. A closer look will be needed.

(iii) *Identification of efficient energy use potential:* The energy intensity is quite high and there is substantial scope for energy efficiency improvement across the economy. The sub-sectors need to be clearly defined.

(iv) *Regulatory reform for energy efficiency, trade and investment:* The need for increased investment flows in the energy sector and also to make non-commercial sources of energy more viable requires the removal of barriers to all forms of energy. Restriction on sale of electricity in the final form from different primary sources is a major barrier for private power production. Energy efficiency improvement also requires removal of market distortions through regulatory reforms.

(v) *Reduced environmental pollutants emission from energy use and production:* The obligation to reduce environment polluting emissions calls for controlling carbon dioxide, sulphur dioxide and nitrous oxide emissions. Such a change will need either a reduction in use of environment polluting energy sources or switching over to alternative fuels. Different possible abatement measures need to be persued and weightage for each of these measures be reflected in terms of cost.

(vi) *Regional Cooperation in energy infrastructure connections:* Regional cooperation in sharing the energy resources will improve availability of energy within the region, provide flexibility, improve the security of supply and enhance the efficiency and effectiveness of the whole energy supply system. Recent efforts of the Institution of Engineers (India) as a Member of the Federation of Engineering Institutions of South and Central Asia, FEISCA; and as Indian Member Committee of the World Energy Council have promoted this concept on a regional basis in South Asia.

Conclusion

As the awareness for environmental concerns is spreading, it is also necessary to build these concerns by internalising them into the cost structure of the energy supply and demand. In a major exercise conducted by the Indian Member Committee of the World Energy Council, under the guidance of the Ministry of Power, Government of India and the Central Electricity Authority, a consensus was arrived to set up a South Asia Regional Energy Centre in India to become the focal point for quality energy and other relevant data availability for the region and for developing such strategies and models that will ensure availability and security of energy supply in South Asia region. Setting up of such a Centre has to be the first priority in the energy sector. It is urgent to consider the structural set-up and the functional set-up.

REFERENCES

Chandra, Harish, Environment Indicators, National Workshop on Environment Statistics, Goa, 12-13 January, 1999.

Goel, R.S., Envoronmental Impact Assessment for Water Resources Projects, Symposium on Earth Sciences in Environmental Assessment and Management, Lucknow, 9-11 April 1999.

IIASA-WEC Report on Global Energy Perspectives, 1998.

India's Energy Scene, National Paper, 17th Congress of WEC, Houston, 13-17 September, 1998.

Kulshrestha, A.C. *et al.*, System of National Accounts and Environmental Accounting, National Workshop on Environment Statistics, Goa, 12-13 January, 1999.

Maggo, J.N., Analysis of Wood Energy Issues in National Level Five Year Development Planning Exercises - Annual Convention of Bio Energy Society of India, New Delhi, 14-15 December, 1995.

Moudgal, S., Environmental Legislation in India, Symposium on earth Sciences in Environmental Assessment and Management, Lucknow, 9-11 April, 1998.

Narasimhan, R.L., Development of Environment Statistics, National Workshop on Environment Statistics, Goa, 12-13 January, 1999.

Parikh, Jyoti, *et al.*, Climate Change and India's Energy Options, IGIDR, Discussion Paper No. 76, May 1992.

Parikh, Jyoti, *et al.*, Consumption Patterns: The Driving Force of Environmental Stress, IGIDR, October 1991.

Sengupta, R.P., Energy Modelling for India : Towards a Policy for Commercial Energy, Planning Commission of India, New Delhi, 1993.

Voluntary Exchange of Information, Report from Government of India, 4th Session of UN Commission on Sustainable Development, 1996.

Yokobori, Keichi, Asia Pacific Energy Prospects for 2010–17th Congress of WEC, Houston, 13-17 September, 1999.

5

INFORMATION TECHNOLOGY FOR THE ENERGY SECTOR

Introduction

World energy demand has been growing at an alarming rate for nearly 50 years. This unprecedented growth is the direct result of rapid improvement in living standards in the developing countries as well as relatively peaceful relations between most of the world's nations. Such rapid growth in energy consumption poses many challenges. Two of the most critical issues centre around:

(i) The availability of reliable and affordable energy resources, and
(ii) The environmental impacts from the consumption of these energy resources.

'Information Technology' holds the key to development in the next millennium. Energy sector development is no exception to the same and will be dictated largely by the revolution due to information technology. Energy comprises 5 per cent of the world's US $ 30 trillion GDP; the more efficient use of those energy sources, the greater efficiency can be injected into the remaining 95 per cent of total world output. Efficiency will be built into energy sector only through judicious use of information technology.

A large scale, optimised and integrated management system would be vital for the future power utilities. Opportunities for the energy industry in the 21st Century are as fascinating as the revolutionary changes taking

---

Paper presented at the 13th Indian Engineering Congress, organised by The Institution of Engineers (India) at Chandigarh during April 25-26, 1999.

place in many growth businesses. Barrier-free markets are attracting new competitors who have an extensive global presence, sharp mass marketing skills and highly respected brands. These entrepreneurs are uninhibited by industry standards and practices. Thus, they think more creatively about products and services that are attractive to consumers.

Two Grand Energy Transitions in the Past

The World Energy Council has analysed development of energy sector over the 19th and 20th Centuries. It has been observed that prior to the Industrial Revolution the energy system relied on harnessing, natural energy flows and animal and human power to provide required energy services in the form of heat, light, and work. Power densities and availability were constrained by site-specific factors, with mechanical energy sources limited to draft animals, water, and windmills. The only form of energy conversion was from chemical energy to heat and light through burning fuelwood, for example, or tallow candles. Energy consumption typically did not exceed 0.5 tons oil equivalent (toe) per capital per year.

Two 'grand transitions' have since shaped structural changes in the energy system at all levels. The first was initiated with a radical technological end-use innovation: the steam engine powered by coal. The steam cycle represented the first conversion of fossil energy sources into work; it allowed the provision of energy services to be site independent, as coal could be transported and stored as needed; and it permitted power densities previously only possible in exceptional locations of abundant hydropower. Stationary steam engines were first introduced for lifting water from coal mines, thereby facilitating increased coal production. Later, they provided stationary power for what was to become an entirely new form of organizing production: the factory system. Mobile steam engines, on locomotives and steamships, enabled the first transport revolution, as railway networks were extended to even the most remote locations and ships were converted from sail to steam. Characteristic energy consumption levels during the 'steam age', approximately the mid-19th century in England, were about 2 toe per capita per year. By the turn of the 20th century, coal had replaced traditional non-fossil energy sources and supplied virtually all the primary energy needs of industrialized countries.

The second grand transition was the greatly increased diversification of both energy end-use technologies and energy supply sources. Perhaps the most important single innovation was the introduction of electricity as the first energy carrier that could be easily converted to light, heat, or work at the point of end use. A second key innovation was the internal combustion engine, which revolutionized individual and collective mobility through the use of cars, buses, and aircraft. Like the transition triggered by the steam engine, this 'diversification transition' was led by technological innovations in energy end use, such as the electric light bulb, the electric motor, the internal combustion engine, and the aircraft. However, changes in energy supply have been equally far-reaching. In particular, oil emerged from its place as an expensive curiosity at the end of the 19th century to occupy the dominant global position, where it has remained for the past 30 years.

These two grand transitions show the changing shares of different primary energy sources in the global energy supply, including the long transition away from traditional renewable energy forms toward fossil fuels; the emergence and culmination of coal, which supplied close to two-thirds of global energy needs by the eve of First World War; the introduction of oil, and later, natural gas, first as a by-product of oil production and then as an energy carrier in its own right; the peak in the market share of oil in the 1970s; and finally a reduction in the dynamics of change in the primary energy supply structure during the past two decades. This reduced dynamism may be partly due to the increased regulatory interest received by the energy sector in recent decades, partly due to oil's attractiveness for the transportation sector, where demand has risen steadily, and partly due to the delayed switch in power generation away from coal and oil to natural gas.

The two grand transitions have made possible far-reaching structural changes in employment, the spatial division of labour, and international trade. These are associated with modernizing traditional economic and social structures, and include the following, in particular:

Industrialization

Employment and value generation move progressively away from agriculture toward industry and manufacturing in particular. Subsequently there are structural shift away from 'smoke-stack' industries toward services and industries characterized by the increasing importance of information generation and handling.

Urbanization

Spatially, urbanization implies a drastic relocation from rural to urban residence, employment and economic activities. Socially, urbanization entails deep changes in the social fabric and the emergence of new values and lifestyles. Economically, urbanization is driven by the large and diverse economic opportunities that the agglomeration of mainly enterprises and consumers entails. As a result, the largest portion of a country's gross domestic product (GDP) is generated in urban areas and agglomerations. As an extreme case, it is estimated that 80 per cent of the GDP of Indonesia is generated in the national capital. For energy, urbanization imposes strict quality requirements for the energy carriers used, including higher power density and cleanliness.

Important changes in energy demand and supply associated with these two grand transitions include the following:

Commercial Energy

There has been a transition from noncommercial to commercial energy forms that reflects the structural economic shift from agriculture to industry, the related monetization of the economy, and increasing urbanization.

Increasing Energy "Quality"

There has also been a transition from direct use of solid energy forms, such as traditional biomass and coal, to liquids and grid-dependent energy forms that are more flexible, more convenient and cleaner. This is a function of three major underlying trends in the 'quality ladder' of different energy currencies:

— Industrial processes and technologies are becoming ever more complex, with increasing requirements for ease of handling, storability, continuity, and flexible availability of energy supplies.
— Requirements for convenience increase with rising levels of affluence, so that in addition to price there is an increasing 'convenience premium' determining fuel choices in residential and commercial end uses.
— Both of these trends lead to higher 'form value' (quality) of the energy currencies that are at the interface between energy supply and demand, favouring flexible, clean energy forms such as electricity, gas , and ultimately hydrogen. Put succinctly: one cannot operate a computer directly with fuelwood or coal.

Decreasing Energy Intensity

Although per capita energy needs have increased with economic development, the specific energy needs per unit of economic activity have decreased. This ratio is referred to as energy intensity.

Next Energy Transition—To be Influenced by Information Technology

The 17th Congress of the World Energy Council held in Houston during September 1998, emphasised on the role of information technology in the energy sector supply and demand. The Congress concluded in this aspect as follows:

'As the energy industry is restructured globally, new businesses and alliances are emerging in the upstream production and generation activities as well as in energy distribution. These developments are improving efficiencies in locating and extracting energy resources from deep and remote parts of the globe. New distribution alliances are coupling energy with efficient technologies to provide services customers want, including transportation, lighting, air conditioning and refrigeration services. Information technology is making markets more efficient, resource production less speculative and costly, and the delivery and monitoring of energy use more effective, enfranchising customers to make more intelligent choices.'

Preparations for the 21st Century indicate that the rate at which technology is maturing and impacting our world is more rapid and more profound than any time in recent history. Much of this is being caused by substantial advances in computer and communications technology:

(i) Four million people used the Internet in 1994, but over 100 million do so today (the corresponding figure for India is dismal low of 90,000 only). Ford, USA, uses links with six countries to exchange 3-D model and complex design images. Boeing 777 was created entirely for master 3-D models on networks linked by computer engineers. Already, communication technology is moving swiftly past the Internet to a public multi-media network where high band width application, high intelligence information devices and portability will provide access to any information at any location and for every change the way we run our business and conduct our personal lives.

(ii) A number of examples can be picked up on how this shift to an information and knowledge economy is impacting the oil and gas industry. Horizontal drilling was coined a "remarkable leap" 15-20 years ago, but in late April this year (1998), the US oil industry responded with equal enthusiasm to "Downhole Factory" that will combine fibre optics, robotics, artificial intelligence and other technology to refine oil or convert natural gas into electric power within the wellbore.

Kathryn Kranhold has described the impact of information technology on energy use sector in the Wall Street Journal of September 14, 1998, as follows:

'It's 3 O'Clock in the morning in the year 2003, and suddenly you are awakened by a loud spinning noise. It's your programmable washing machine starting up on its own. You roll over and shut your eyes, knowing that you are saving a few pennies on your electricity bill by doing your laundry when power is cheap.

The next morning, in the office, your utility calls with bad news: Your basement is flooded. How do they know? Because a new electricity meter you have installed also gives warnings of frozen pipes, flooded basements and burglars.

This is the world the electric companies have dreamed up for you: a deregulated market where power suppliers compete against one another by offering technological innovations and new choices for customers.

But it's a long way from here to there. In fact, as the U.S. begins its journey into electricity deregulation, probably only one thing can be predicted with any certainty: It's going to be awfully confusing a couple of years.

For customers, it will mean having to shop for electricity, in the same way they shop around for cars, computers or long-distance telephone service. They'll have to compare prices, services and reputations, and make up their own minds about who gets to supply power to them. For utilities, it will mean transforming themselves into companies that actually have to sell a product, not just deliver it. And for investors, it will mean taking a closer look at management's strategies and portfolios to determine what companies have the right mix.'

Deregulation and Unbundling

The main message for the 21st century change is simple, powerful and fast. Deregulation in the US and other developed countries, has significantly changed telecommunication, energy and other industries in the last ten years. The US electricity industry is rapidly restructuring. Energy companies are developing new strategies to attract and retain customers, divesting of local assets and acquiring distant ones, implementing new trading and market businesses, restructuring balance sheets and expanding into international markets. And in Europe, the European Union has developed into an amalgamation to become one economic community with individual states and with coordinating time tables for opening industrial markets for electricity and gas competition.

The infrastructure system of developing countries are on the threshold of significant and rapid transformations. Ten years ago, these changes were non-existent, ten years from now , most of these countries will have evolved into full-fledged, competitive market that are also regionally integrated. By the time, the next energy transition in 2020 takes place the energy markets in the developing countries will be quite competitively geared up to enter the global markets. The process of unbundling is being promoted in developing countries. The Latin American economy is boosting power demand by approximately 6 per cent per year. India has opened up its infrastructure sector for privatization, the electrical power demand is increasing @ of 7.5 - 10 per cent per year and its total power needs over the next ten years are estimated at over US $120 billion.

Optimisation of Integrated Energy Supply

If integrated energy supplies are to become the norm, then the optimisation process would, need to be continued by escalating it on to higher distribution level in order to open up new options for bottom-up energy supply optimised according to economic and ecological criteria. The basis for this is already to be seen in the electricity supply industry.

Integrated communication - initially only necessary to optimise the energy supply system will also pave the way for billing management, more flexible incentive based tariff structures and at the same time serve as a basis for the new utilities to explore new business opportunities.

Building an enhanced domestic services are just two examples of where opportunity knocks. Once a foothold has been established inside a building or home with a service connection, the prospects for offering other communications related services will be enhanced: TV, ISD and telephony, the Internet, home banking, building services management and security are just a few of the potential areas of opportunity. The extent of these new and challenging opportunities has without doubt yet to be fully realised.

Dr. Ing Rainer Bitsch of Siemens Power Transmission and Distribution Group has developed a vision for an energy supply concept for the next millennium. Three different scenarios are envisaged, partly in competition with one and another but at the same time complementing each other in a meaningful way. Urban areas and industrial regions with power needs in the order of gigawatts in future will continue to be served by a conventional centralised grid with large generating capacity and powerful energy interfaces - just as they are today in Europe. In urban concentrations, the problem of route shortages could be solved and system voltages reduced by the use of super-conducting cables. Dynamic energy storage and power conditioners will ensure contractually guaranteed quality at the inter-faces to sensitive or demanding consumers. Inadmissibly high short-circuit currents will be prevented by super-conducting current meters.

At the other end of the supply spectrum are domestic dwellings with installed loads of the order of a few kilowatts. In future, these could be served by combined heat and power systems in the form of packaged co-generation units, batteries of fuel cells, a micro turbine running generation or if incident sun-light levels are sufficiently high, photovoltaic generation.

Between these two extremes of demand are the collective MW consumers, who live in blocks of flats and works in small industrial firms. This is the level at which, locally available energy resources can be optimised with the minimum environmentally adverse impact. This will be made possible by the use of intelligent and decentralised energy management system. Such system would provide key data to the next level in an operational hierarchy and when additional supply and demand values are taken into consideration will come closer than ever before in achieving the optimised supply of energy.

The network control systems of the future will tell the users as to what is available from various alternative sources at any given time. Assuming similar terms and conditions, consumers tend to prefer local supply and services over national or international ones.

These network control systems will direct local networks made up of small scale energy generating systems (probably based on wind, solar, water, biomass or gas) and the means of storage (possibly in the form of a battery) and industrial and private consumers. Diesel power sub-stations or other fossil generating facilities already in existence can also be integrated in the same way. In addition to electrical energy, a biomass power plant can also generate heat, which is extracted and supplied to consumers for drying, processing or heating.

Innovative Approaches for Customer Relations

Utilities in the United States are also trying to convey a more dynamic, innovative image, spending millions of dollars on advertising and image makeovers. Many have dumped stodgy names that marked them as utilities in favour of new-age identities that sound good to focus groups. When Public Service Company of Colorado merged with Southwestern Public Service Company last year, the combined company called itself New Centuries Energy. San Diego Gas & Electric created a holding company it called Enova Corporation, and then, after merging with the parent of Southern California Gas, renamed itself Sempra Energy Corporation.

Utilities are also changing the way they think about their customers. Unlike in the monopoly days, keeping customers' business now means keeping them happy. So the companies are spending millions of dollars on educating their employees about customer service. Some are coaching employees to call consumers "customers" instead of "ratepayers".

Some utilities are offering extra services-telephone, home security and cable services, for instance - alongside power. Workers for Mid American Energy Holdings Company, Des Moines, Iowa, are supposed to ask their customers, 'Seed or sop?' after tearing up a lawn to install new equipment. Meter readers and service crews may also issue a $25 voucher on the spot if a customer has been inconvenienced.

These companies will also be pushing innovative products your way, such as high-tech electric meters and more energy-efficient appliances, to complement their service. One new type of meter, developed by Enron and CellNet Data System Inc of San Carlos, California, would tell you exactly how much electricity you use every hour, so you can figure out what you can expect to pay at the end of the month. A rival meter developed by ET Communications, San Francisco, is the one that would also warn you of a flooded basement.

Conclusion

Today's technology calls for efficiency gains in the production, transportation and distribution of energy. It is not the generation technology alone that is undergoing a revolution, but there has been significant achievement in tools that help consumers use energy more efficiently. Information technology will be the tool for the third energy transition and the informed customer will dictate energy supply options.

REFERENCES

Bitsch, Dr. Rainer, *Energy Concepts for the Millennium,* Power Economics, October, 1998.

Kranhold, Kathryn, Impact of Information Technology on Energy Sector Use, *Wall Street Journal,* September 14, 1998.

Lay, Kenneth L., Change and Innovation: The Evolving Energy Industry, 17th Congress WEC, Houston, September, 1998.

Nakicenovic, Nebojsa, *Global Energy Perspectives, WEC/IIASA Study,* Cambridge University Press, 1998.

Ybema, Remko, Prospects for Technological Progress and Utilization of Renewable Resources, 17th Congress WEC, Houston, September, 1998.

6

QUALITY ENERGY SERVICES NECESSARY FOR THE NEXT MILLENNIUM

Introduction

The policy of liberalisation in India has brought into focus the quality concept. While in the early 90s a number of industries publicised that they have the ISO 9000 certification, that is no longer the case. The important industrial players consider that ISO 9000 certification is an essential requirement for them to keep abreast of others in this competitive world. A timeframe of 6 - 7 years is not too long for bringing around a revolution in a country which is accused of being bureaucratic in approach. It is a reflection of the maturity and understanding of industrialists and businessmen in this country, who out of their natural wisdom realised the importance of quality. The Institute of Directors created a facility for industrialists (of all categories) to access quality certification since 1992. There can be absolutely no doubt that the country would be marching into the new millennium with quality as its hallmark. India having entered global energy market can attain quality energy services by following global strategies in the energy sector in the next decade and beyond. The Paper deals with various issues to show as to how technology, financing and pricing will be important in developing options for ensuring quality energy services in the coming years.

Global Observations on Energy Supply

The largest, and the last event, of this century in the Energy Sector was

---

Paper presented at the World Quality Congress 1999 organised by the World Quality Council at Bombay during February, 1999.

the 17th Congress of the World Energy Council, organised in Houston during September 13-18, 1998. The U.S. President, Mr. Bill Clinton in his message said, 'In today's global community, we realise more than ever the importance of energy efficiency to our security, productivity and prosperity. We are fortunate to live in a time of great promise for the development of sustainable energy sources. New technologies and growing public awareness of the need for energy efficiency offers us an unprecedented opportunity for progress in our efforts to manage wisely, the world's energy resources'.

Over 8,000 participants from all over the world deliberated for a week to find out as to how best the technologies could provide them the options for meeting the energy provisions for atleast over the next 50 years. The WEC Congress considered the ways in which energy technologies could :

— advance energy supply for those currently without commercial energy services.
— raise efficiency, both in energy provision and use so as to mitigate adverse environmental impacts, including potential climate change, to move towards a sustainable world and to improve the competitiveness of economies.
— reduce local and regional pollution through the wider use of currently available technologies in transport and power generation, and efficient end-use equipment.
— address climate change concerns by deploying more efficient fossil fuel technologies and by using non-fossil fuel energy resources.
— place emphasis on renewable energy resources to supplement conventional resources.

The deliberations identified that the technologies will play an important role, but, in themselves will not be sufficient to fix all problems. Even development of technologies and their implementation will call for major policy and financial support from the government. Therefore, institutional and policy changes will be required. Options will have to be made attractive for international funding at right places. Public awareness and public attitudes and behaviour will need a major shift. Public curiosity will not be satisfied by information dissemination alone but they will be keen to receive answers for the queries which come up from time to time.

The participants focused their attention on the direction that the energy markets could take in the next 50 years. There was recognition that old certainties and paradigms are giving way to new markets, challenges, organisations and ways of doing things. The major industries or the major current players expressed their concern that in over a decade a large number of major energy players may be reduced to producers and sellers but not the leading industrialists in the energy sector. Future lies in the hands of such industries who have linked themselves with information technology and are willing to change with time. The expectation was that a large number of smaller players, who build up their outfit with information technology back-up will control energy sector.

Global Energy Perspectives: 2050 and Beyond

World Energy Needs will Increase

The World Energy Council, WEC, and the International Institute of Applied Systems Analysis, IIASA undertook a five year study to project the future energy demands to the year 2050 and beyond. This study describes three cases of alternative energy futures that diverse into a total of six scenarios and their implications for 11 world regions. The three cases have been classified as: Case A presents a future of impressive technological improvement and consequent high economic growth. Case B describes a future which is ambitious, though perhaps more realistic, technological improvements, and consequently more intermediate economic growth. Case C presents an ecologically driven future. It includes both substantial technological progress and unprecedented international cooperation centered explicitly on environmental protection and international equity.

World population is expected to double by the middle of the 21st century, and economic development needs to continue, particularly in the South. According to the scenarios of this study, this results in a 3 to 5-fold increase by 2100. At that stage, the per capita income in most of the currently developing countries will have reached and surpassed the levels of today's developed countries. Disparities are likely to persist, and despite rapid economic development, adequate energy services may not be available to everyone, even in 100 years. Nonetheless, the distinction between 'developed' and 'developing' countries in today's sense will no longer be appropriate. Primary and final energy use will grow much less than the demand for energy services due to improvements in energy intensities. The study expects a 1.5 to 3-fold increase in primary energy requirements by 2050, and a 2 to 5-fold increase by 2100.

Energy Intensities will Improve Significantly

As individual technologies improve, conversion processes and end-use devices progress along their learning curves, and as inefficient technologies are retired in favour of more efficient ones, the amount of primary energy needed per unit of economic output, the energy intensity decreases. Other things being equal, the faster the economic growth, the shorter the obsolescence time; the higher the turnover of capital, the greater the energy intensity improvement. Combined with the economic growth patterns of different scenarios, the average global reductions in energy intensity range between 0.8 per cent per year and 1.4 per cent per year. The South Asia region with limited domestic resources and having experienced rapid economic growth and energy intensity improvement can have further rapid improvements which will be more of a necessity to sustain rapid growth rather than a possibility.

Resource Availability will not be a Major Global Constraint

The resource scarcity perceived in the 1970s did not occur as originally assumed. With technological and economic development, estimates of the ultimate available energy resource base will continue to increase. A variety of assumptions about the timing and extent of new discoveries of fossil energy reserves and resources, and about improvements in the economics of their recoverability, are reflected in the range of scenarios considered by the study. All, however, indicate that economic development over the next century will not be a constraint by geological resources. Regional shortages and price increases can occur, due to the unequal distribution of fossil resources, but globally they are not a constraint. Environment concerns, financing, and technological needs appear more likely sources of future limits.

Quality of Energy Services and forms will increasingly shape Future Energy Systems

The regional reviews strongly endorsed the global conclusion that the historical trend by which final energy forms become cleaner and more grid-dependent as affluence grows can continue for a wide range of possible future primary energy structures. Solid fuels such as traditional biomass and coal in residential and commercial applications are gradually phased out in all scenarios. Grid-dependent fuels such as electricity, natural gas, and eventually also hydrogen are phased in. Although the speed of the transition varies, the trend everywhere - in global and regional perspectives alike - is toward cleaner, more convenient, and more flexible end-use fuels, The consistency between the regional reviews and the global analysis in their emphasis on the structure of final energy use is particularly important, as this feature of the study represents a conceptual and methodological advance over traditional scenario studies focused solely on energy supply perspectives.

Energy End-use Patterns will Converge, even as Energy Supply Structures Diverge

The six scenarios indicate that the historical drive toward even more convenient, flexible, and clean fuels reaching the consumer can continue for a wide range of possible future energy supply structures. In all scenarios, there is a shift toward electricity and toward higher-quality fuels such as natural gas, oil products, methanol, and, eventually, hydrogen. In contrast to these converging trends, primary energy supply structures diverge, particularly after 2020. Fossil sources continue to provide most of the world's energy well into the next century but to a varying extent across the scenarios. There is a shift away from non-commercial and most unsustainable uses of biomass, and direct uses of coal virtually disappear. Sustainable uses of renewables including modern biomass come to hold a prominent place in all scenarios. They reach the consumer end as electricity, liquids, or gases, rather than as solids.

Technological Change will be Critical for Future Energy Systems

Technological change drives productivity growth and economic development and is crucial under all scenarios. Technological markets are expected to grow substantially creating continuing opportunities for improved, adapted and new technologies. But progress has a price. RD&D of new energy technologies and the accumulation of experience in niche markets required up-front expenditures of money and effort. These are increasingly viewed as too high a price to pay in liberalised markets where the maximisation of short term share holder value generally takes precedence over longer term, socio-economic development and environmental protection. Yet, it is the RD&D investments. Of the next few decades, that will shape the technology options available after 2020.

A robust hedging strategy focuses on generic technology at the interface between energy supply and end-use, including gas turbines, fuel cells and photovoltaics.

Technological change increasingly crosses national borders. However, efforts to transfer technology have sometimes failed because successful technology diffusion requires high local expertise. Infrastructure need to be in place, labour must be educated, and local RD&D capabilities are required for the necessary local adaptations, modifications, and extensions of new technologies that make the difference between success and failure. For this reason, capacity building is a resilient strategy for preparing for a transition away from fossil fuels to a more diversified energy supply system. If indigenous technological capabilities are not nurtured right from the outset, however, sunk cost in infrastructure human capital make it progressively more difficult to change course.

Technological change is directly linked to the quality of energy services required. The investment in RD&D will also be proportionate to the service.

Capital requirements will Present Major Challenges for all Energy Strategies

For all scenarios the capital requirements of the energy sector are large, but not infeasible. Although investment requirements expand more slowly than overall economic growth, the energy sector will have to raise an increasing fraction of its capital from the private sector. It will face stiffer competition and return on investment criteria than it has in the past. Moreover, the greatest investment needs are in the new developing world, where current trends in the availability of both, international development capital and private investment capital are not auspicious.

It is realised that the market pricing only make an investment attractive if there is market that can pay the price and the poor and developing countries do not yet constitute such a market. How quickly they become empowered energy consumers will depend more on the success of policies to reduce poverty than on energy policies.

Global Energy Industry Forming New Alliances

The restructuring of the global energy industry is resulting in new businesses and alliances appearing in the upstream production and generation activities as well as in the downstream energy distribution activities. These developments are improving efficiencies in locating and extracting energy resources from deep and remote parts of the globe. New distribution alliances are coupling energy with efficient technologies to provide services customers want, including transportation, lighting, air-conditioning and refrigeration also. Information technologies are making markets more efficient, resource production less speculative and costly, and the delivery and monitoring of energy use more effective, while enfranchising customers to make more intelligent choices. Overall one can say that the energy industry is moving globally towards more quality production and supply system and environmentally cleaner regime.

Quality Energy Services in Future

The energy system is basically service driven, from consumer to producer, while energy flows are resource and conversion-process driven, from producer to consumer. Various studies have indicated the need to consider energy end use and energy supply simultaneously, both from an analytical and policy perspective. In addition to prices and quantities, energy quality matters increasingly. Quality considerations include convenience, flexibility, efficiency and environmental cleanliness. The energy system is end use driven. Under market liberalisation, special competitive advantages will arise for those companies prepared to deliver a full range of energy services beyond just fuels and electricity.

Following three transformations have been predicted:

1. All the three cases for projection of energy demand to year 2050 reflect a continuing pervasive shift from energy used in its original form, such as traditional direct uses of coal and biomass, to elaborate systems of energy conversion and delivery. This shift is projected for all the three scenarios leading to ever more sophisticated energy systems and higher quality energy carriers.
2. A second profound transformation is the increasing degree to which energy is delivered by dedicated transport systems, such as pipelines and networks. This development enhances trade possibilities and promotes similar end-use patterns across regions with fundamentally different primary energy supply structures.
3. The third profound transformation is that the changes in final energy patterns reflect the changes in economic structure

presented in the scenarios. As incomes increase, the share of transport and residential/commercial applications also increases.

These converging final energy patterns yield substantial quality improvements in the final energy (and resulting energy services) delivered to the consumer.

The efficiency of final energy use improves as the final energy carrier portfolio changes in the direction of higher quality fuels. The effect is an improvement via interfuel substitution of 20 to 30 percent. The actual end use efficiency gains are of course much larger, as they are mostly driven by technological change in end-use devices. The main point is that the more efficient end-use devices will also require higher quality fuels.

Conclusion

Quality has long been the preserve of the market controlled economy of the developed countries. Recent advancements in the developing countries have also raised awareness about the quality energy services. Global advance into the next millennium will call for stringent technological advancements to ensure uninterrupted, quality energy services for global economic and social development. But, that will call for appropriate pricing of energy sources; funding of frontier technologies; and poverty alleviation measures in developing countries—all that appear outside quality considerations' but crucial for ensuring quality energy services.

REFERENCES

Annual Report 1997-98, Ministry of Power, Government of India.

Annual Report, Ministry of Non-Conventional Energy Sources 1997-98, Government of India.

Conclusion & Recommendations, International Conference on Financing of the Energy Sector, Indian Member Committee of WEC, New Delhi, July 1998.

Economic Survey of India 1999, Government of India.

Nakicenovic, Nebojsa, *Global Energy Perspectives-WEC/IIASA Study*, Cambridge University Press, 1998.

National Paper, India's Energy Scenario 1998, Presented at the 17th Congress of WEC at Houston, September 13-18, 1998.

7

ROLE OF CONSULTING ORGANISATIONS IN POWER SECTOR

Introduction

Energy sector is a major element of the national infrastructure and has received priority attention during the 90s. Lack of finance has been the main reason for the government to raise power generation capacities. The government accepted professional global advice to liberalise and privatise. These two terms have been given utmost importance. Various steps have been experimented during the last decade to take action conforming broadly to this terminology. This has created awareness in the private and foreign investors about the need for investment in country's primary infrastructure sector. This awareness could not be converted into action due to political uncertainties and fluctuating currency exchange rates.

Liberalisation of Power Sector: What does that Involve?

Liberalisation and privatisation have come to stay. But, a proper understanding of the same is desirable.

Observations at the 17th Congress of the World Energy Congress, at Houston, in September 1998, give indication to future preferences. The consequences of liberalisation in the energy sector have been thoroughly analysed. There has been a general overall support for liberalisation but that needs to be properly understood. Some specific remarks can be observed as follows:

Paper presented at the National Seminar on 'Privatisation Policy - Successes & Failures', organised by Consulting Engineers Association of India, at New Delhi, on February 19, 1999.

(i) The first observation is whether with ongoing liberalisation energy will become simply a 'commodity' as any other good or service. Energy differs from a typical commodity in atleast two essential points: conventional energy resources, in common with others, such as minerals, are finite and cannot be reproduced. Also, energy is indispensable for the social, economic and industrial development of any country. Due to this strategic value no government in practice would leave this precious sector to market forces only.

(ii) The second observation mentions that liberalisation must not be regarded as an end in itself. Rather, it is an instrument to achieve certain objectives essential for a specific sector. For the energy sector, these objectives are to ensure a reliable, long term supply of energy in sufficient quantity at affordable prices from processes that are compatible with prudent environmental standard.

Measured against these criteria, liberalisation of the energy sector can provide very positive development - in releasing new synergies, raising efficiency, reducing cost, promoting new investments, cutting prices and last but not the least in introducing a new mental approach. All this will help to make the overall supply more stable. There is a strong possibility of exceeding the environmental target and allowing extensive cross subsidies of one kind or another to be removed.

A number of conclusions have been drawn regarding the impact of liberalisation of the energy sector:

— First, neither haste nor hesitation will bring positive results - what is needed is clear strategic thinking, transparent and consistent frameworks and both a determined and pragmatic approach to problem solving.

— Second, the needs, the conditions and the methods for energy sector recognise that liberalisation differ greatly from country to country. Consequently, liberalisation is neither the only answer or the solution of all problems, nor is there a single way, a single model of how to achieve liberalisation. Liberalisation is not an easy process and there will be winners and losers.

In case of India the situation greatly differs from state to state as the condition of State Electricity Boards is at great variance: some of them have opted for early liberalisation and have shown the path successfully, others have not even initiated action.

Government has recently taken a number of measures in the power sector. Some of these can be better understood by studying the process of reforms that have been gone through in the power sector.

In this general background an assessment has to be made of the role of consulting organisations. The starting point obviously will be the policy measures and implementation strategies of the government.

Power Sector Reforms

Functional Reforms (Investment Promotion)

Reforms in the power sector were initiated in 1991. In order to make the climate for investment congenial to the entrepreneurial measures for deregulation and devolution of power to utilities, direct negotiation with the entrepreneurs has been encouraged. Regulations relating to fuel supply have been liberalised so as to attract a variety of fuels to become supportive to power generation based on economy of supply. Legislative measures were effected to support congeniality in the commercial climate in order to make tariffs acceptable to the consumer and encouraging to the entrepreneurs. The clearance procedures have been rationalised. Time for techno-economic clearance by CEA has been reduced to 5-6 months. Project clearance based on competitive bidding is made mandatory and the methods of clearance are being rationalised.

Structural Reforms

In structural terms, power sector utilities are endeavouring to unbundle the existing vertically integrated monopoly into separate entities for generation, transmission and distribution for rationalising their management. Whereas efforts are afoot to maintain conceptual uniformity in unbundling, multi-ownership for generation and distribution and continuation of government monopoly in the transmission would be the near term approach in the initial stage of reforms. Utilities are currently debating the possible structures. The existing system (State Electricity Board structure) in some utilities may continue in the transition period with functional and managerial improvements (as a few utilities have done creditably well even in the existing dispensation of vertically integrated structures).

Setting up Electricity Regulatory Commission

In order to provide a congenial and confidence building climate and to promote healthy competition and reasonable and transparent tariff making arrangements, the Central Electricity Regulatory Commission (CERC) has been set up at the central level. The CERC, has been charged with the following responsibilities:

(i) to regulate the tariff of generating companies owned or controlled by the central government;
(ii) to regulate the inter-state transmission including tariff of the transmission entities;
(iii) to regulate the inter-state bulk sale of power;
(iv) to aid and advice the central government in formulation of the tariff policy etc.

The process of unbundling of generation, transmission and distribution, establishment of regulatory mechanisms and readjustments in the existing structures are at different stages of implementation in different federal utility systems. In Orissa, the State Electricity Regulatory Commission has since been set up. Regulatory systems would be on anvil in Rajasthan and Gujarat. On the other hand, in states like Andhra Pradesh, Maharashtra and Tamil Nadu, attention is being paid for readjusting the existing structures. In other states, the choice of the unbundled systems is at different stages of discussion and debate. Other areas that would come under the purview of the regulatory commissions (at the state level) will include regulation of the working of licensing, protection of the interest of the consumers, ensuring reasonability in the operational efficiency and pricing of electricity.

Environmental Management Reforms

Environmental management and its internalisation in the project conceptualisation, operation and maintenance has also emerged as one of the key factor in the reforms process in the National Agenda. The changing economic structure, in line with the global economic and commercial environment for liberalisation and deregulation, has also been given due weight in the planning for environmental upkeep related to development activities.

India's power sector has to work in unison with the global power sector development. Following issues on the power sector will have a bearing on actions in India.

- The first issue is that the information technology is now taking over the traditional practices and major companies with an edge in information technology will have the control of global markets.
- The second important issue is that the global availability of conventional energy sources will not be a constraint but the supply lines may lengthen in certain cases. This will call for effective and optimum use of locally and regionally available energy sources.
- The third important aspect is the transport or transmission of energy from one place to another at an economic cost. Regional Grid for South Asia, will therefore, take precedence over other activities.
- The fourth important issue is the allocation for site specific research, development and demonstration activities. This research will also focus on environmental issues. Such actions and activities alone will give a long term sustainability to the energy systems.
- The fifth important issue is that energy is now to be dealt as a service rather than a commodity. It will have to be supplied in the form and quality acceptable to the consumer. The future will be much more demanding on the quality of energy supply.
- The sixth important issue is that locally available renewable sources - mainly biomass and hydro sources of energy will have to be tapped for small power generation units and mega scale plants will have to be developed for centralised base load applications.
- The seventh important issue is that the international cooperation and collaboration across the board can improve effectiveness of performance of such systems.

Priority Areas for Consulting Organisations

While dealing with the priority areas for the consulting organisations and their role this paper will refer only to such consultancy areas that cover the technological aspect. Strategic and management consulting approaches are not being dealt with in this case. The priority areas have been identified considering that liberalisation and privatisation will continue; and efforts will be made to keep pace with the global action.

The priority areas for the consulting organisations can be identified as follows:

— Identification of suitable technologies and processes for transfer and their adoption into power plants and transmission system.
— Identification of measures necessary for quality power.
— Identification of energy sources, development of project proposal for identified sites; development of work plan; and its implementation.
— Design, planning and implementation of transmission grid. (This includes the sophisticated electronics that goes into stabilizing the transmission grids).
— Development of suitable energy conservation measures and energy efficiency improvement on the supply side as well as the demand side.
— Development and introduction of clean technology for future environmental conservation.
— Development of joint and collaborative arrangement with leading international consulting organisations so as to ensure appropriate adoption of technologies suitable to local conditions (as the foreign organisations transfer their experience only).
— Develop a network of consulting organisations in the South Asia Region so that hydro electricity, gas and other energy sources can be suitably exploited and shared to mutual benefit, by utilisation of their own technological capabilities.

Conclusion

The impact of globalisation and de-regulation, backed by telecommunication revolution is tremendously reducing the distance between homes and businesses. It is expected that with this effect, the monopoly of electricity distribution from centralised sources will lose importance, and the need for high quality, high reliability power near the home will increase (distributed generation will increase). As long as India is dependent on coal as its primary source of energy, mega power projects will continue to dominate the scene as they are financially attractive. Whatever form of energy supply and conversion is adopted the control of the existing public utilities and public companies will continue well into the next century. These companies will be slow to move. Public funding into the private industries in the power sector will grow at a very slow and gradual pace. That was observed even in UK when privatisation of energy industry was undertaken. In other words, any remarkable change in the set up of existing public companies may not take place at a fast pace as that will disturb the general structure of these energy companies. Out of their compulsion to face the challenge from emerging private and foreign companies, the public sector energy companies will have to upgrade their technologies and skills. The only possibility for them will be to out-source some of their activities to technologically advanced enterprising consulting organisations. Flexibility of operation and availability of latest knowledge with them will be the most striking feature.

The major trends leading into the 21st Century are creating enormous opportunities for visionary consulting organisations who can tap the vast Indian markets and the South Asian regional markets. As the acceptability of some of the major Indian companies will improve in the international markets and they become multinational, the Indian consulting companies will have greater chance of entering into the international arena.

At present the consultancy organisations can be the most effective link for converting the political decisions into practical programmes through the existing energy companies and the emerging energy companies.

REFERENCES

Annual Report 1997-98, Ministry of Power, Government of India.

Annual Report, Ministry of Non-Conventional Energy Sources 1997-98, Government of Nakicenovic, Nebojsa, *Global Energy Perspectives, WEC/IIASA Study*, Cambridge University Press, 1998.

Conclusion & Recommendations, International Conference on Financing of the Energy Sector, Indian Member Committee of WEC, New Delhi, July 1998.

Economic Survey of India 1999, Government of India.

Eighth Five Year Plan (1992-97), Planning Commission, Government of India.

Maggo, J.N., *R&D Needs for Energy Sector*, Department of Science & Technology, Government of India, New Delhi, October 1996.

National Paper, India's Energy Scenario 1998, Presented at the 17th Congress of WEC at Houston, September 13-18, 1998.

8

SOUTH ASIA REGIONAL COOPERATION IN ENERGY SECTOR

Introduction

The need for development of infrastructure in the South Asia Region is important to provide better quality of life to the common man. FEISCA has appropriately taken up the challenge of identifying infrastructure development activities that can be supported by the engineering fraternity, and motivating its members to work in that direction.

Characteristics of Situation in South Asia

Demographic and economic characteristics of South Asia Region (SA) and energy consumption patterns can be summarised as follows:

(i) SA includes all the SAARC countries.
(ii) India, Pakistan and Bangladesh represent three of the most populous countries.
(iii) India represents 80 per cent of GDP and 75 per cent of region's population.
(iv) In 1990 the Region had :
- 20 per cent of global population.
- 2 per cent of gross world production (GWP)
- 3.2 per cent of global energy related net carbon emissions.
- Less than 10 per cent of world average per capita income @ $ 334.
(v) Economic growth during 1990-98 is 5.3 per cent over the last

Paper presented at the Regional Workshop on 'Infrastructure Development' organised by the Federation of Engineering Institutions of South and Central Asia, at Kathmandu, on March 3, 1998.

decade.

(vi) Commercial energy use during 1990-96 grew at 4.5 per cent per year, from 223 mtoe to 291 mtoe.
In case of India the commercial energy use grew from 177.61 mtoe to 249.38 mtoe which is roughly 80 per cent of the total commercial energy consumption in the region.

(vii) Commercial energy intensity decreased by 0.9 per cent per year in the same period.

The more rapid the economic growth, the faster the energy intensity will improve as old equipment is replaced by new equipment and the economic structure shifts towards less energy intensive activities.

WEC Study Highlights for Future Energy Scene

The World Energy Council has released, in 1998, its observations on future perspectives of energy supply and consumption patterns in South Asia. Major issues of concern are as follows:

(i) South Asia will grow to first place by the end of next century from 7th place among the 11 regions in terms of energy consumption in 1990.

(ii) Hydro-carbon imports will be the major concern ranked highest in the South Asia Regional Review.

(iii) Institutional deficiencies will have to be overcome. A shift from subsidies and flat rates to segmented market pricing will be necessary; and losses and leakages will have to be reduced and efficiencies improved.

(iv) The problem of 'market exclusion' rather than of 'market distortion' is important.

(iv) Local environmental impact are a high priority concern. Except for the ecologically driven scenario the carbon dioxide and sulphur dioxide emissions are likely to increase.

Relevance to the Host Country (Nepal)

Nepal, the host for the Conference, holds an important position in South Asia. Use of locally available energy resources has been an important element of Nepal's economic development. Biomass and hydropower are the two indigenous energy resources used to fulfill the energy needs. Efforts to further exploit hydro potential can certainly create a situation where Nepal can become a net exporter of energy in the most desired form of electricity. Difficult geographic situation, environmental restrictions and lack of funding have created a situation where this potential could not be fully exploited. ESCAP and SAARC have indicated in their studies that hydro electricity can be most appropriately shared amongst Nepal, India, Bangladesh and Bhutan. FEISCA, at its Conference in August 1998 at Dhaka, took initiative to identify working mechanism whereby Regional Grid could be developed on commercial lines; and facilitate energy sharing.

The local consumption of electricity is low due to lack of purchasing power of the people. The electricity service is estimated to be available to about 15 per cent of the population. Country is, however, endowed with a vast hydropower potential estimated at 83,000 MW, out of which about 25,000 MW have been investigated at different levels of details. Nepal Electricity Authority (NEA), a government owned electric utility, is the main player in the power sector in Nepal. NEA is vested with the responsibility for the construction, ownership and operation of electricity generation, transmission and distribution facilities through the Kingdom of Nepal. NEA generation capacity is characterized by a preponderance of hydro generation, accounting for 84.19 per cent of 299.81 MW of installed capacity.

NEA was established in 1985 through a merger of several institutions with an objective of bringing about efficiency and coordination in the power sector. NEA is under the jurisdiction of Ministry of Water Resources.

Nepal views efficient exploitation of enormous hydropower potential as an important national priority, both for the domestic market as well as for the export to India. In recent years, a number of developments have taken place in the fulfilment of this objective, notable among them being the signing of Power Trade Agreement between Nepal and India. This is expected to open up doors for large scale exchange of power between Nepal and India. Additionally, private companies have also shown interest in developing hydroelectric projects with a view to exporting power to India.

Crucial Link for Regional Cooperation

South Asia Regional Cooperation holds the key for development and cost effective sharing of energy resources in the region. FEISCA had organised a Regional Workshop on Power Sector at Dhaka during August 5-7, 1998. The deliberations proved effective in motivating governments and industries in this region to launch initiative for closer cooperation. This outcome was also encouraging for the multilateral funding agencies like the World Bank and Asian Development Bank. They offered to support a larger number of projects of regional nature as they offered better rate of return and reduced risk of recovery of their funds.

Transmission and transportation of electric power and other energy sources has always been a major constraint. Power Grid Corporation of India Limited, PGCIL, has developed an ambitious plan of investing about Rs. 18,000 crores (US $ 4 billions) during the Ninth Plan period (1997-2002) for creating and strengthening transmission links, leading to the formation of a National Grid. This investment will develop a national grid to transmit 10,000 MW of power by the end of 2007.

Power Grid Corporation has also taken decision to develop a Regional Grid so as to be in a position to evacuate power from Nepal, Bhutan, Bangladesh and Pakistan. Mr. R.P. Singh, Chairman-cum-Managing Director, PGCIL, in an interview to *The Times of India,* published on February 23, 1999 said, 'There is growing realisation worldwide that trading of natural resources is necessary for their optimum utilisation. In the case for exchange of power, the reliability of supply increases, there is economy in operations, and there can be mutual support during emergencies.

Negotiations are going on for transfer of power from Pakistan to India. Transmission links will have to be built, initially on the Punjab border, and in the long term, on the Rajasthan border. We have a small inter-connection with Bhutan. There are plans to build 400 KV lines to transfer power from the proposed hydro projects in Bhutan through the chicken-neck area to Eastern India.

With Nepal, too, a number of small inter-connections exist. Transmission lines of 765/400 KV are envisaged for evacuation of power from upcoming hydro-electricity projects in Nepal. With Bangladesh, two inter-connections are envisaged, initially for an exchange of 150 MW'.

Nepal has planned for High Voltage Transmission Network. That will ensure stability in the power system and guarantee transmission of quality power. Nepal Electricity Authority deals with India basically at 132 KV power system. Nepal gets lot more power in 33 KV and also in 11 KV system from India. The Nepal/India Power Exchange Committee has decided 50 MW of power exchange in 132 KV in Duhabi-Kataiya system, which in the later stage will go up to 150 MW. The Nepal Electricity Authority has emphasised on setting up Load Dispatch Centres (LDC), to meet the future expansion needs of the transmission network.

Support for Regional Cooperation at the Highest Level of Governance

In informal discussion the Prime Minister of Nepal (at the inauguration of the 6th National Convention of Nepal Engineers' Association, at Kathmandu, on December 3, 1998) also emphasised on the need for India and Nepal to discuss the water sharing treaty and have a new look, so that higher hydro-electricity potential could be tapped in Nepal.

This is to be seen as a major contribution of FEISCA. Their platform at Dhaka and Kathmandu helped in crystallising the approach. The dialogue has not reached an end but needs many more rounds of talks at various levels. FEISCA's contribution becomes significant when one views the firm commitment from the governments of Bangladesh and Nepal to involve in programmes of mutual benefit through South Asia Regional Cooperation. The Institution of Engineers(I) obviously has contributed its bit.

The 17th Congress of the World Energy Council at Houston in September, 1998 was another occasion for launching a regional cooperation initiative between India and Pakistan. (Though the Prime Ministers of India and Pakistan at the SAARC Meet at Male in early 1998 showed interest in entering a cooperation programme to evacuate 4000 MW of power from Pakistan to India, no further concrete proposal was put forth. A stage had reached when such a cooperation move was put in the cold storage). WEC organised a Round Table on 'Developing Countries' Energy Challenges: A Discussion with Ministers'. Hon'ble Mr. P.R. Kumaramanglam, the Minister for Power, Government of India and Hon'ble Chaudhary N.A. Khan, the Minister for Petroleum & Natural Resources, Government of Pakistan were on the Panel. During the discussion a suggestion emerged that the two Ministers and their officials meet and start talks on cooperation on energy sharing. Hurried official action followed and the two Ministers deliberated alongwith their senior officials. It is understood that such an action was appreciated by respective Prime Ministers. Two more rounds of talks have been organised to transfer 4,000 MW capacity from Pakistan to India. It is expected that with the involvement of energy companies an acceptable arrangement will be established.

Promising Categories for Regional Cooperation

These actions are indicative of the reality that regional cooperation in the energy sector is high on the political agenda of SAARC member countries. If that is true then action on following issues need be taken up immediately to establish/promote regional cooperation:

(i) Sharing Energy Resources.
(ii) Sharing Manpower and Know-how.
(iii) Sharing Investments.
(iv) Sharing Environmental Control Measures.
(v) Energy/Industry Collaboration.
(vi) Sharing Experience of Professional Societies under the Umbrella of FEISCA.

Details of these categories are given below.

(i) Sharing Energy Resources

The 17th Congress of WEC projected that primary sources of energy are in abundance to meet the future global needs well into the next century. However, their geographic location will create major problems to some of the developing countries. South Asia could be seriously affected under both the criterion. It is, therefore, necessary to analytically assess the locally available energy resources and work out a strategy for sharing the same on a regional basis. Such an approach can be mutually beneficial and also ensure long term sustainability of energy supply.

These resources can be shared mainly on commercial line through the market mechanism. Geopolitics of the region will always have a bearing on national decision making, but the operational mechanism should be dictated by the market forces and commercial viability.

Specific sources of energy that can be shared are dealt with in the following paragraphs:

(a) *Hydro-electricity:* India, as a major partner in South Asia, can play a pivotal role in developing a sound power sector in the region. Prospects of jointly developing a number of new large sized projects in Nepal (e.g. Pancheswar - 6,000 MW; and West Seti - 750 MW) and in Bhutan (e.g. Tala - 2,000 MW) are under consideration from where a major portion of power would be supplied to India. Presently India and Nepal are sharing hydro-electricity on a limited scale and Bhutan supplies bulk power to India from Chukha project. Other sites can also be developed.

(b) *Natural Gas:* Bangladesh can supply excess natural gas to India or generate power from the same and transmit that to India.

(c) *Coal:* India can supply larger quantities of coal to Nepal for its domestic application.

(d) *Regional Grid:* Power Grid Corporation of India Ltd has vast network of transmission grid in India. It has plans to develop a Regional Grid connection amongst Bangladesh, Bhutan, Nepal and Pakistan. Completion of the regional grid with appropriate support systems (like the Load Dispatch Centres in Nepal) will effect smooth transmission of power.

(ii) Sharing Manpower and Know-how

(a) *Sharing Technical Know-how on Different Energy Sources - Mainly the Renewable Sources of Energy:* Liberalisation and privatisation process has been introduced in India, Bangladesh and Pakistan; and is also being introduced in Nepal and Sri Lanka. This has involved the private sector in a specific enhanced fashion, whereby giving access to technical know-how from market sources. Recently absorbed technologies can be appropriately repackaged for sharing in the region. Similarly, the locally developed renewable energy technologies can be mutually shared in the region. The terms for such sharing will have both the instruments - one supported by the governments and the other based on market forces.

(b) *Sharing Technical Manpower in the Region:* Human Resource in the energy sector for policy making, planning, project implementation, management and operation of energy systems can be mutually shared. The terms for such a sharing will depend on market forces.

(c) *Sharing Institutional Mechanism:* Institutional changes are required to formulate appropriate policies to effectively implement them and to take steps to coordinate actions both at the national and regional levels. This often lacks coordination amongst different agencies responsible for planning energy. Pressure groups in respective countries need to be developed by involving energy industries, investment banks, energy utilities and government organisations.

(d) *Setting up a Regional Energy Centre for South Asia:* Regional sharing of resources - physical, manpower and financial - needs a new mechanism to be developed. Regional cooperation in terms of projects that help in exploitation of resources in one country and its sharing through trade with the others can be developed on the basis of mutual investment sharing and product sharing.

A Regional Energy Centre for South Asia needs to be set up with the support of governments in the region. Activities of the Centre may be guided by a governing council having representatives from different countries.

(iii) Sharing Investment

(a) *Sharing Investment in Plants and Machinery:* Such an approach will be beneficial in more than one country and will also build investor confidence. Attracting finance, both domestic and external, in the energy sector requires government's commitment to ensure commercialisation of energy sector supported by government policies for unbundling, sharing and management of risks that are applicable to all potential investors in the energy sector. Investments in generation, transmission and utilization systems can effectively enhance the overall availability of energy supply and will also ensure cost effective use of investments.

(b) *Setting up a Regional Fund:* A regional mechanism in the form of a Energy Fund for South Asia should be set up to pool resources from multilateral, bilateral and private donor agencies to support bankable energy projects and to take action to mobilise investments for the establishment of energy projects. This will help in building a comprehensive energy infrastructure in South Asia as the Fund will be able to justify allocations to the donor agencies and also ensure timely assessment and evaluation of the progress of various projects.

(iv) Sharing Environmental Control Measures

(a) *Sharing Information from Environmental Monitoring Institutions:* Setting up an institution for environmental monitoring and control is a time consuming and a costly financial proposition. Some of the existing institutions in the region can appropriately develop packages that will help in monitoring and controlling the quality of air and water, which is directly affected by energy use. Such an approach will fulfill obligations under the Kyoto Protocol of the Framework Convention on Climate Change; and will also ensure sustainability of energy supply and systems in an environment friendly fashion.

(b) *Setting up Environmental Monitoring Stations:* A constant vigil on environmental conditions will be necessary to promote use of appropriate energy system. This will call for low cost environment monitoring equipment and processes at a large number of points.

(v) Energy Industry Collaboration

Energy industry collaboration is crucial for rejuvenation of energy supplies (in the final form) to the users. Apex industry focal points should take initiative for strengthening regional cooperative programmes by creating infrastructure and manpower skills to promote energy security in the region. This arrangement will support the bilateral and the multilateral programmes of various governments. Involvement of the industry will also bring in efficiency and flexibility in operation.

(vi) Sharing Experience of Professional Societies under the Umbrella of FEISCA

A number of professional institutions are operational in different countries and have linkage to engineering institutions. It would be appropriate for FEISCA to develop a mechanism so that experience of these professional societies can be coordinated and shared under its umbrella.

Conclusion

Regional Cooperation in energy sector will not emerge in isolation, but will be a part of overall development strategy in South Asia. Professional inputs are crucial in preparing a case for the government to consider regional cooperation seriously. Various elements identified in this Paper cannot be dealt by any one professional group but needs inputs from different sources. FEISCA is a 'networking mechanism' that can bring engineering community on one platform; and also motivate respective governments to take collective decision of mutual interest in the energy sector.

REFERENCES

Annual Report 1997-98, Ministry of Power, Government of India.

Annual Report, 1997-98, Ministry of Non-Conventional Energy Sources, Government of India.

Conclusions & Recommendations, International Conference on Financing of the Energy Sector, Indian Member Committee of WEC, New Delhi, July 1998.

Eighth Five Year Plan (1992-97), Planning Commission, Government of India.

Economic Survey of India 1999, Government of India.

Maggo, J.N., R&D Needs for Energy Sector, Department of Science & Technology, Government of India, New Delhi, October 1996.

Nakicenovic, Nebojsa, *Global Energy Perspectives, WEC/IIASA Study,* Cambridge University Press, 1998.

National Paper, India's Energy Scenario 1998, Presented at the 17th Congress of WEC at Houston, September 13-18, 1998.

9

ENVIRONMENT FRIENDLY ENERGY SUPPLY THROUGH BIOMASS IN THE NEXT MILLENNIUM

Introduction

Sustainable energy supply in a clean environment is the biggest challenge for the mankind. Moving a step forward from providing end of the pipe cleaning up technologies to green and clean technologies in the energy sector requires multi-disciplinary and multi-sectoral approaches. Recent actions have focused on clean technologies and cleaner sources of energy. Experts have argued that starting with an environmentally benign source of energy would be the key to attaining sustainability. Discussions at UNCED (1992) and Kyoto Conference of Parties of the UN Framework Convention on Climate Change (1996) have gone a long way on emphasising on cleaner sources of energy. The Kyoto Protocol is still an open document awaiting ratification by December 2000. Details of Clean Development Mechanism (CDM) have still to be worked out by January 1, 2000. The Kyoto Protocol has set in motion a major revolution in the field of development and every country has been persuaded to think in terms of cleaner approaches in the energy sector.

The political determination of the Government of Malaysia to support energy technologies and utilisation of environment friendly sources of energy was reflected in the declaration by the Prime Minister on June 5, 1999 (while inaugurating the World Renewable Energy Congress) that the renewables will be the fifth source while planning for energy. For a country with abundant oil reserves this decision is obviously influenced by environmental concern.

Paper presented at the World Renewable Energy Congress 99, organised by the Ministry of Energy, Communication and Multi Media, Government of Malaysia in Kuala Lumpur, Malaysia during 8-11 June, 1999.

Global energy policy makers have to focus attention on the poor living condition of over 2 billion people in the developing world, who have remained isolated from electricity and commercial sources of energy. A large proportion of them are also projected not to get commercial sources of energy in the near future. Are these communities really deprived of all kinds of energy sources and lack the socio-economic development opportunities? Closer scrutiny shows that even the rural poor in the developing countries without electricity and commercial sources of energy have the means to meet their minimum sustenance energy needs in an appropriate fashion through biomass use. The World Bank projected in 1992 a figure of 2 billion who were undergoing commercial energy poverty. The global advice was to develop market based energy supply systems and ensure accessibility to such markets by the rural poor. The World Energy Council in its report titled "Global Energy Perspectives" (1998) projected that such a market mechanism was useful only for those who have a role in market economy. These 2 billion are poor and have no purchasing capacity to take advantage of the market based system. All such population survive on biomass fuel, which under a number of situations in the developing world are available free of cost. The biofuels are, therefore, termed the poormen's fuel. The accessibility to commercial sources of energy through market mechanism has not become a reality. The poor have no role to play in the market mechanism due to lack of purchasing capacity. Ultimately they are a case of 'market exclusion'. The reality is that biomass energy is the fuel for the poor. Biomass fuels are used by 3 billion or half the global population to cook their food every day.

Biomass is contributing as a source of energy in the developed and the developing world. The South Asia, with a population of over 1.2 billion, is largely dependent on biomass energy for meeting the energy needs in member countries. Contribution of such traditional biomass energy in the total energy consumption varies from 33 per cent in India to 92 per cent in Nepal.

Sustainable Energy Supply through Biomass

Sustainable energy supply through locally available resources has been emphasised globally by inter governmental and professional organisations. Use of biomass has figured prominently for the developing countries. Following references are relevant:

UN Conferences

The decade of the 90s witnessed a series of major conferences on global issues. UN Conference on Environment and Development UNCED, at Rio, in 1992; the World Summit for Social Development at Copenhagen, in 1995; and the Habitat-II Conference at Istanbul, in 1996 have deliberated on social and environmental goals and objectives for the international community. These conferences dealt with energy as it impacts sustainable development. In addition, two major global conventions negotiated in this decade also concern biomass energy. United Nations Framework Convention on Climate Change and the Convention to Combat Desertification also addressed energy issues as they relate to climate change and land-degradation. These and the associated platforms of action from the UN Conference have been unanimous in articulating the need for changes in the current approaches to energy to include more attention to the sustainability of energy systems and the ability to provide energy services that can support the achievement of socio-economic and environmental goals and objectives.

WEC Congress

World Energy Council, WEC, the premier global organisation of energy experts/agencies also focused attention on energy sustainability at the 14th Congress at Montreal, in 1989. Subsequently, all its triennial Congresses in 1992, 1995 and 1998 focused on sustainable energy supply. The consideration of supply options for the 2 billion poor without access to commercial sources of energy in the developing countries has attracted attention of global energy players. Biomass and other renewable sources have come centre state.

Overview on Energy Use

Over the past 30 years, the world's energy requirement has risen considerably. In 1960, the world used 3.3 Gtoe (gigatonnes of oil equivalent) of energy, and in 1990, 8.7 Gtoe, an increase of 163 per cent or an average increase of 3.3 per cent per annum.

Energy use per capita is even more varied. In 1990 the majority of the world's population used less than 1.2 toe (tonnes oil equivalent) of energy per capita (indeed many use less than a third), the energy rich (in North America) use more than 7 toe per head. Under the commercial energy, the per capita energy consumption in India is still one of the lowest in the world. It recorded 123.5 kgoe (kilograms oil equivalent) during 1990-91. However, if non-commercial energy consumption is also included, the per capita consumption goes up to 265.4 kgoe. The per capita electricity consumption in India recorded 253 kwh. The growth in demand has been fastest in the developing countries. In the past decade, it has risen in developing countries by 49 per cent, and in the developed countries by only 14 per cent. Although most of the energy is still consumed by the developed industrialised countries, the proportion accounted for by the developing countries can be expected to continue to increase.

Future Biomass Energy Use

Global Energy Perspective or World Energy Council

WEC Report: "Energy for Tomorrow's World" (1992)

The WEC Commission in 1992 developed three Global Energy Cases, each representing a different assumption in terms of economic development, energy efficiency and environmental impact:

(i) The first case 'A' is referred as the Reference Case (REF). It is in essence the same as developed by the WEC for the Montreal Congress 1989, updated by subsequent experience, and considering "business-as-usual" scenario'.

(ii) The second case 'B' is the Enhanced Economic Development Case (EED) which assumed a somewhat higher rate of economic growth in the developing countries than the Reference Case but with somewhat reduced energy efficiency; and

(iii) The third case 'C' is of the Ecologically Driven Case (ED) which assumes the same economic growth as in the Reference Case, but with a sharper improvement in energy efficiency.

WEC Report: Global Energy Perspectives to 2050 and Beyond (1995)

WEC in its report "Global Energy Perspectives to 2050 and Beyond", published in 1995 further refined these cases into six different scenarios.

Case 'A' includes three scenarios addressing key developments in energy supply. They vary principally and that in the future they envision coal on one side, and nuclear and renewables on the other. In the A1 scenario, there is a large future availability of oil and gas resources.

Dominance of oil and gas is perpetuated to the end of the 21st century. At the other end of the spectrum, the A2 scenario assumes oil and gas resources as scarce, resulting in a massive return to coal. Finally, in the A3 scenario, rapid technological change in nuclear and renewable energy technologies results in a phase out of fossil fuels due to economic reasons rather than resource scarcity.

Case 'B' has only one scenario and reflects a pragmatic approach. It manages to fulfil the development aspirations of the South, but less uniformly, and at a slower pace. For regions such as Africa, progress is painfully slow. Overall, economic growth rates to 2020 are more modest than in Case 'B' of Energy for Tomorrow's World (1992).

In Case 'C' the nuclear energy is at crossroads. Two scenarios are included. They both meet the carbon dioxide emission ceiling in 2020, but they describe two very different paths that nuclear power might take.

WEC Report : Global Energy Perspectives (1998)

WEC published its report 'Global Energy Perspectives' 1998 to refine the six cases defined in 1995. Projections have taken into account the impact of environmental issues. A projection to 2100 is also included.

WEC Projections for Energy Use (1992)

Future energy supply in the three cases in the foreseeable future is projected as given in Table 9.1 below:

TABLE 9.1: Future Energy Demand (in Gtoe)

1990	2020		
	REF	EED	ED
8.7	13.3	17.2	11.2

REF: Reference Case; EED: Enhanced Economic Development Case.

ED: Ecologically Driven Case.

Source: WEC's Energy for Tomorrow's World (1992).

Table: 9. 2 reflects the possible "Global Energy Supply Mix" for the three cases in the year 2020 as compared to the corresponding energy mix in 1990.

TABLE 9.2: Energy Mix in Global Fuel Use (in Gtoe)

Energy form	1990	2020		
		REF	EED	ED
Coal	2.3	3..2	4.8	2.1
Oil	2.8	3.7	4.6	2.7
Gas	1.7	2.8	3.5	2.3
Nuclear	0.4	0.8	1.0	0.7
Hydro	0.5	1.0	1.2	0.9
New Renewable	0.2	0.5	1.3	1.5
Traditional* Biomass	0.8	1.3	1.2	1.0
Total	**8.7**	**13.3**	**17.2@**	**11.2**

Source: WEC's Energy for Tomorrow's World (1992).

* Only fuelwood, agro residues and cow dung and does not include biomass recoverable by new technologies.

@ Rounding off the figures show a deviation of 0.1 in the total.

WEC Projections of Biomass Energy Use (1998)

Whereas the above are the indicative projections for the biomass use prepared by the WEC-IIASA, there are other projections also. D.O. Hall et al have presented these as given in Table 9.3.

TABLE 9.3 : The Role of Biomass in Future Global Energy Use (EJ)

Year Scenario		2025	2050	2100
Shell	(1996)	85	200-220	—
IPCC	(1996)	72	280	320
Greenpeace	(1993)	114	181	—
Johansson *et al*	(1993)	145	206	—
WEC	(1993)	59	94-157	132 - 215
Dessus *et al*	(1992)	135	—	—
Lashof and Tirpak	(1991)	130	215	—

Source: Hall & Scrase - WEC's World Resources Handbook (1998).

FAO Biomass Use Projection (1998)

WEC projections do not go into details on the sources and applications in the biomass sector. FAO collects, analyses and presents detailed clarification of energy use under different categories.

FAO's Forestry Department recently established a Wood Energy Information System (WEIS) which collected existing data through the implementation of Regional Studies (Wood Energy Today for Tomorrow-WETT) (1997). These regional studies made a detailed analysis of existing information available in FAOSTAT and other important wood energy databases, such as UN-Energy Statistics, IEA, EUROSTAT, UN-ECE, OLADE, RWEDP etc. The information and data collected and compiled has been organized according to the new set of terminology and definitions prepared by FAO (Unified Wood Energy Terminology, FAO, 1997). The new terms and definitions adopted contribute to better understanding of the role of different woodfuel sources, the classification of woodfuels by categories and a clearer definition of the major users of rural and urban areas.

Table 9.4 summarises the woodfuel consumption in the main regions of the world, divided between developed and developing countries. It includes data from 147 countries arranged in line with the FAO classification already used in other documents, such as the State of the World's Forests (FAO, 1997).

The total production of wood in 1995 reached approx 3,900 million cum, out of which 2,300 million cum (Table 9.4) have been used for woodfuels. This means that approx 60% of the world's total wood removals from forests and non-forests lands have been used for energy purposes.

FAO has projected that in global terms, woodfuels produced either directly or indirectly from forest resources are more than half of the total forest removals. Or, in other words, taking into account the volume of wood consumed, energy is the main application of forests. However, FAO's Regional Wood Energy Development Programme for Asia, operating in 16 developing countries in Asia, collected data for their member-countries and projected that 2/3rd of woodfuels come from non-forest resources (1997).

World production of fuelwood for energy use in 1996 is estimated to be about 1.9 billion cum or 1.4 billion tonnes (FAO 1998). An estimated additional amount equivalent to some 0.3 billion tonnes of wood is recovered in recycle for energy use. The total of 1.7 billion tonnes of fuelwood is equivalent to 550 million tonnes of oil. Compared

TABLE 9.4: Total Woodfuel Consumption by Regions (1995)

Region	1000 cum	PJ	Ratio: woodfuel/ all energy	Ratio woodfuels/ all wood uses
Developing countries total	1,763,262	17,633	0.15	0.80
Total tropical	1,368,439	13,684	0.26	0.84
Total non-tropical	394,822	3,952	0.06	0.65
Africa	486,248	4,862	0.35	0.89
tropical	464,077	4,641	0.75	0.91
non-tropical	22,171	222	0.03	0.53
Asia	1,002,846	10,028	0.12	0.81
tropical	654,221	6,542	0.23	0.85
non-tropical	348,625	3,486	0.07	0.70
Oceania	5,804	58	0.52	0.56
Latin America and Caribbean	268,364	2,684	0.12	0.66
tropical	244,338	2,443	0.13	0.69
non-tropical	24,027	240	0.07	0.48
Developed countries total	536,754	5,368	0.02	0.31
Europe, Israel and Turkey	194,653	1,947	0.03	0.33
Former USSR	42,585	426	0.01	0.27
Canada and USA	272,438	2,724	0.03	0.29
Australia, N. Zealand and	22,079	271	0.01	0.29
Japan				0.36
World	2,300,016	23,000	0.07	0.59

Source: FAOSTAT, United Nations Energy Yearbook 1997, Wood Energy Today for Tomorrow Studies (FAO, 1997).

Cum: Cubic metres.

to 9 Giga Tonnes of oil equivalent (Gtoe) of primary energy supply in 1990 it stands at 6-7%. (Woodfuels exclude biomass other than woodfuels).

Most of it (76%) is used in developing countries. The balance of (24%) fuelwood used in developed countries (approximately 5400 PJ) represents only 2% of their total energy consumption. 44% of the total woodfuel consumed in the world is used in Asian countries.

Its share is somewhat less than that of nuclear, but it is substantially the largest source of renewable energy, providing more than twice the contribution of hydro electricity worldwide. Even in Europe and North America the woodfuel contribution is on par with that of hydro-electric output.

The importance of fuelwood in the global energy supply varies drastically between regions and also within a region. In Europe and North America, taking into account re-cycling in industry, wood contributes about 3%. In the industrial countries of Asia and the Pacific, the contribution is about 1.5% and in the former Soviet Union it is about 2-3%. In the developing countries, with greater rural populations, the proportion of the rural energy supply met by fuelwood is higher. In Asia and Latin America 10-15% and in Africa it is 33% of the energy requirements.

While only 30 per cent of the wood produced in developed countries is used for energy (33% in Europe and 29% in North America), in developing countries that amount reaches 80%. In fact a high proportion of the total wood consumption is used for woodfuels in Africa, Asia and Latin America where woodfuels account for 89 per cent, 81% and 66 % of their respective total consumption, (as presented in Table 9.4). Over the past 50 years the volume of wood used for energy has approximately trebled (an average growth of 2% per annum). The trend in industrialised countries upto the mid 1970s was for decline. This was reversed with the energy price shift and there has since been some growth, particularly when residue recovery and waste recycling for energy use are taken into account.

However, divergence in use pattern within a continent is much greater. In case of South Asia the dependence on woodfuel, on an average, is 37% (1996). However, at the country level, the variations are marked. This kind of variation has crept in due to large consumption of fossil energy in India. In case of Nepal, woodfuel contribute 92% of the total energy supply, whereas in India its contribution is about 33% of the total primary energy supply.

In developed countries, wood energy contribution also vary considerably from country to country. In Europe, for instance, the UK, Belgium and Germany, there is relatively little use of woodfuels while in countries such as Finland, Sweden and Austria, forest energy provides up to 17% of their demand. France, as well as Germany, deserve special mention because forest energy provides a meagre 4% of its national energy consumption but the 329 PJ of woodfuels used is similar to the amount used in Sweden (326 PJ) which represents more than 16% of the total energy consumption. It should also be noted that many countries have no or relatively little information on wood energy, especially those from the former USSR.

In the EU biomass supplied about 45 mtoe (3%) of its primary energy (in 1996), but with wide variations, e.g. 12% in Austria, 18% in Sweden, and 23% in Finland. By the year 2010 about 135 mtoe (8.5%) of the primary energy could be derived from biomass, representing an additional 90 mtoe (EU White Paper, 1997). An important characteristic is that biomass energy in the rich countries has evolved/is evolving against a background where forests and woodlands are generally under-utilized and where people are able to choose from a range of energy options and have the purchasing power to buy them. In the European Union, however, it is perhaps surprising to learn that most forest energy is still used by households where woodfuels represent around 60% of the total wood energy consumed.

In developing countries biomass is the main source of energy for many artisan and cottage industries such as baking, brewing, textile manufacture, tobacco, coffee and tea curing, fish-smoking, brick-making, etc. For example, in Asia, rural industries account for at least 20% of the region's wood energy consumption (FAO 1996). Charcoal is used in heavy industry - pig iron, steel, cement and metallurgy. Brazil consumes over 6 million tonnes of charcoal annually in heavy industry.

Contribution of Woodfuel Types

Availability of a number of sources of woodfuel for each specific location contradict the old perception that fuelwood and charcoal are removed from forests and are thus the main causes of deforestation, FAO's Regional Wood Energy Development Programme in Asia. RWEDP, studies have confirmed the important role played by non-forest sources in woodfuel supply in Asia, such as trees and shrubs from farming lands, irrigation canals, roadsides and marginal lands. RWEDP has relied on reports submitted by respective governments differentiating between fuelwood and other residues.

Table 9.5 shows the share of different woodfuel types derived directly and indirectly from forests, as well as non-forest lands. According to Troserro, (1998), FAO considered difficult to assess the fraction of wood residues used as energy, therefore, all figures recorded in FAOSTAT as woodfuels incorporate wood residues.

Data confirms importance of traditional fuelwood demand, especially in developing countries. On the other hand, generation of energy from recovered black liquor is high in Oceania, North America and Europe, where large pulp and paper industries have heat and power plants fueled with these biofuels. Almost all of their energy needs are met and sometimes surpluses are sold to the public grid.

TABLE 9.5 : Share of Woodfuel Types in Total Wood energy

Region	Woodfuels (mcum equivalent)			
	Fuelwood	Charcoal	Residues	Black liquor
Developing Countries Total	1,533	131	15	34
Africa	445	72	2	3
Asia	859	25	10	12
Oceania	6	0	0	0
Latin America and Caribbean	223	34	2	19
Developed Countries Total	187	8	43	227
Europe, Israel and Turkey	56	2	33	51
Former USSR	32	0	3	8
Canada and USA	96	4	1	146
Australia, N. Zealand and Japan	3	0	6	23
World	1,719	138	58	262

Source: FAOSTAT, 1998.

mcum = million cubic metres.

Potential Global Productivities of Biomass

Competition with food production, and thereby land availability is a major concern for near term fuelwood production.

Productivity on Forest Lands

The world's forests is estimated to be 3.4 billion ha. This is about 27% of the land area. There are in addition some 1.7 billion ha with some tree or woody vegetation cover. The forests are estimated to contain about 440 billion dry tonnes of woody biomass. Allowing addition for the biomass in other wooded land, the total stock is equivalent to some 215 billion tonnes of oil. The forests in industrialized countries (with a quarter of this stock) are stable, with a slight tendency to increase in stock. According to FAO (1997) in the developing countries there is an annual loss of forest, due mainly to conversion of forest to other uses, amounting to about 12 million ha per year or 0.6%.

The annual use of wood for energy represents about 0.3% of the total stock, a rate of use well within the natural growth of forests. The use, however, is not evenly distributed in relation to the stock. The consumption of Brazil, fifth in the world, represents 0.3% of its stock, while the consumption of India, largest in the world, represents 2.5% of its stock.

Some effort is directed at offsetting the loss and depletion of forests. According to D.O. Hall (1995) during the past decade about 40 million ha have been planted in developing countries, doubling the area of plantations. Two-thirds of this planting is on community woodlots, farms and small marginal holdings. These plantings have the function of providing industrial wood and environmental protection as well as wood for energy. The rate of establishment of new plantations is similar in industrialized countries and has been mainly for the supply of industrial wood. While tree planting for energy generation has been carried out, it is currently on a very small scale.

Productivity on Agricultural Lands

In view of the large improvements in productivity levels, farmers in USA and Europe are offered financial packages to leave their lands fallow.

It is estimated that in the United States, about 30 mha have been removed from crop production by 1988. In the European community (12 countries) surplus agricultural land resulting from rising yields and changing agricultural subsidies may reach 15-20 mha by the year 2000, and at least 40 mha in the following century if the productivities continue to increase. These crop lands are already being removed from the intensive farming under "Set-Aside Scheme" of the European Community and "Crop Land Reduction Scheme" of the USA. The agricultural surplus in Europe and North America has attracted large subsidies for lowering the production. According to D.O. Hall (1995), this land represents a significant opportunity to initiate biomass energy production schemes, especially if coupled with the use of agricultural and forestry residues.

This approach is attractive for developed countries.

Productivity on Degraded and Waste Lands

FAO estimated that the agriculture occupied 860 mha in 1982 and 1,477 mha in 1989. At the same time, forest land decreased from 5,200 mha in 1982, to 4,087 mha in 1989, to 3400 mha in 1993. Estimates of degraded and abandoned land generally lie between 700 and 1,000 mha which is equivalent to about half the world's present arable land. The extent of this available land has led the scientists to highlight its potential for use in mitigating the greenhouse effect by managing it to become a carbon sink. Wastelands are regarded as having a good potential for storing carbon in trees due to the relatively low levels of carbon in their soils and vegetation.

This problem is observed in developing countries.

As regards biomass production, productivity values of plantation have been reported to be up to 40 tonnes per hectare for high intensity plantation. However, continuous production facilities at 10-15 tonnes per hectare per year may be quite satisfactory. D.O. Hall has suggested that countries like India should plan their long term policy of biomass plantation on an average yield level of 4 tonnes per hectare per year only.

WEC Projection for Land Use for Biomass Production

To address the issue of possible land-use conflict between biomass for energy and agriculture, WEC have used IIASA's Basic Linked System (BLS) of agricultural models (Fischer *et al.,* 1988, 1994) to calculate the food and agricultural land requirements to the high-growth Case A. The required expansion of agricultural cropland to increase diet and to provide food for an additional five billion people (in 2050) is estimated at 250 million ha, with additional 200 million ha required in developing countries. This compares with 1,440 million ha used by agriculture in 1990.

The required land area for energy biomass production in the high-growth Case A was estimated at 400 to 600 million ha by 2050 and 700 to 1,350 million ha by 2100. In an optimistic scenario (lower bound biomass land requirement), biomass land productivity is assumed to grow to 10 tonnes per ha per year, with two-thirds being produced in dedicated plantations and the remainder being recovered from agricultural residues and from forest management. These assumptions were based on the LESS scenario.

These additional land areas for both agriculture and biomass are available in principle (see Table 9.6) but they stretch future land requirements (and land-use changes) to their ultimate limits. By 2100 agriculture (1,700 million ha) and biomass (700 to 1,350 million ha) could require up to 3,000 million ha—as much land as is currently covered by forests. Land-use conflicts may become a major constraint.

TABLE 9.6 : Land Use in 2050 and 2100 compared with 1990, for Selected World Regions, (in million ha)

	Current Use			Additional Land Use			Potential
				Agric.	Biomass	Biomass	arable
	Forests	Pastures	Agric.	(2050)	(2050)	(2100)	land (b)
ICs (c)	1,770	1,190	670	50	70-100	150-350	—
Africa	630	700	150	95	110-180	140-340	990
Asia (d)	600	880	470	33	160-250	260-340	500
Latin America	890	590	150	72	50-80	140-320	950
DCs (e)	2,120	2,170	770	200	320-510	540-1,000	2,440
World	3,890	3,360	1,440	250	390-610	690-1,350	—

a. For maximum biomass scenarios (A2 and A3), range corresponds to 4 to 6 toe/ha land productivity by 2050 and to 6 to 10 toe/ha by 2100. Lower bounds also assume that 80% (2050) and 67% (2100) of biomass will be produced in plantations. Higher bounds assume 100% plantation biomass.

b. FAO (UN Food and Agriculture Organisation) estimate (Alexandratos, 1995)

c. Industrialized countries, including OECD, EEU and FSU.

d. Including MEA

e. Developing Countries

Environmental Aspects of Fuelwood Consumption

Fuelwood being a renewable entity has the positive impact on the environment during its production cycle. However, during its commercial cycle, it also has certain harmful emissions. The main environmental positive impact is that it is basically carbon dioxide neutral, as long as the rate of harvest equals the rate of re-growth.

The emission of wood energy systems normally depend on the type and scale of the system and its operation. Although in general, small scale systems have the advantage that they require less transport of wood, and therefore, less transport related emissions. Conversion related emissions become lower when the scale of conversion unit increases. The most important emission that can occur are carbon monoxide, polycyclo-aromatic-carbon hydrogens (PACs), nitrous oxide (NOx) and particles.

PACs and carbon monoxide occur as a result of incomplete

combustion of wood and will therefore improve with the efficiency of the conversion unit. Thermal efficiencies in small scale household wood heaters can be around 50% (at lower heating value), but are much lower in less efficient designs. Modern district heating stations in Denmark have efficiencies near 90% (at lower heating value) (Hayden, 1996) which leads to much lower CO and PAC emissions. The situation in the Netherlands can illustrate the CO emission problem of household wood heating systems. Although residential wood heating accounts only about 0.5% of national primary energy consumption, it is responsible for 10% of the Dutch CO emission.

It can be concluded that the present consumption of wood for energy is near to carbon dioxide neutral, but considering the large share of households in wood energy consumption one can pose doubts on the behaviour of these sytems in the field of CO and PAC emissions. This problem may be overcome by improvement of wood heating systems or centralisation of the conversion, which will in general raise conversion efficiencies.

Traditionally biomass energy has often been associated with environmental degradation and health hazards. These are complex and interrelated socio-economic and cultural issues. This view was particularly prevalent in the 1960s and 1970s when the premise was that large-scale deforestation arose as a consequence of over-cutting of forests for fuelwood. The facts indicate that over-cutting was unusually a secondary consequence of a more general failure to care for diminishing forest lands and resources.

Biomass for direct combustion is widely discussed. Production of bio-ethanol for transport vehicles is emerging an application where biomass can substantially avoid carbon emissions.

Automobiles generate more atmospheric pollution than any other human activity and contribute very significantly to the emission of greenhouse gases (notably CO_2), while being the fastest-growing energy consumption sector worldwide. Each year the transportation sector produces 1,800 million tonnes of carbon equivalent emissions (mtCe), or 30% of world carbon emissions. By 2010, the current fleet of 550 million cars may have increased to 1.1 billion and CO_2 emissions will have risen by 65% over the 1990 level, owing to the increased use of oil-derived liquid fuels. The rapid growth of the transport sector will have this major environmental implication.

Use of alternative environmentally-sustainable transportation fuels such as ethanol, biodiesel, etc, offers a good potential for reducing emissions. In Brazil, the large-scale use of ethanol fuel has played a significant role in reducing the level of pollution. For example, the introduction of ethanol-gasoline blends had an immediate impact on the air quality of the large cities, particularly Sao Paulo. Lead additives were gradually reduced as the amount of alcohol in the gasoline was increased and were finally completely eliminated in 1991. The aromatic hydrocarbons, which are particularly toxic, were also eliminated simultaneously. In addition, many other pollutants were also drastically reduced.

Various studies have looked at the potential of reducing pollution through the use of cellulose-based ethanol, which offers the potential of reducing pollution by 80 to 95% in carbon emissions in vehicles. Difiglio (1997) has estimated that in 2025 the use of 40 billion gallons (154.4 billion liters) of cellulose ethanol in the USA would reduce emissions by 100 mtCe, and the total cumulative GHG reduction at 930 mtCe. In Brazil, Moreira and Goldemberg (1997) have estimated that the net contribution of sugarcane amounts to 12.7 mtCe/year. Over a time span of 100 years 330 tC/ha can be avoided in carbon emissions, showing that in the long term ethanol is very effective for decreasing CO_2 emissions.

Role of Energy Plantation in Carbon Sequestration

Sustainably grown biomass is nearly carbon dioxide neutral, depending on production and conversion methods. If biofuel production is to be regarded as a carbon sink, the total amount of carbon stored per hectare under the bioenergy crop must be greater than the level of stored carbon in the vegetation previously on that land. Also, where biofuels are used as substitutes for fossil fuels, the biofuel can be regarded as a sink in terms of the avoided carbon dioxide emission that would have been derived from fossil fuels.

The most clear benefit in carbon sequestration terms would be if the plantation was established in a desert with an existing standing stock of almost 0 tC/ha. Both the above and below-ground level carbon would be raised significantly compared to the existing level. However, if the standing stock of the previous vegetation was greater than that of the new plantation, then a net reduction in the standing C levels would result. If the plantation biomass were to be used as a fossil fuel substitute (or long-lived produce), a new reduction to atmospheric carbon dioxide emissions would occur after a given time period. The higher the levels of vegetation, prior to the plantation, the longer it would take for fuel substitution benefits to reduce the amount of carbon dioxide initially released. This results from the relatively large amounts of the carbon dioxide which are released during the initial clearance through harvesting and transport losses.

Marland G. and Marland S. (1992) in 'Should We store Carbon in Trees' have explored the use of plantation for a fossil fuel substitution plantation-based model (1992). They conclude that the three most important factors in assessing whether biomass energy plantation are effective carbon sinks are as follows:

(i) Carbon inventory of the natural vegetation;
(ii) Productivity of the plantation; and
(iii) Time perspective adopted.

Biofuels provide an opportunity to avoid fossil-fuel-derived carbon dioxide emissions. However, the present intensive agricultural methods can be used only for the production of liquid fuels, e.g. diesel from rape seed, if the associated residue production is used for energy. Thus, the use of carbon-flows and energy 'output : input' ratio to monitor net carbon dioxide production must be integral to the strategies for energy provision.

Experts believe that at good productivities (referring to 6t carbon/ ha/year equivalent to about 12 oven dry t/ha/year biomass) the reforestation of all available land could theoretically remove about 5 GtC from the atmosphere per year over the next 40 years. Present (1998) gross emissions of carbon dioxide into the atmosphere are around 7-8 GtC/year of which 3-4 GtC appears as the atmospheric build-up in the levels of carbon dioxide. (The gross emissions in 1990 amount to 6.23 GtC). Sequestration programme could, therefore, be regarded only as a temporary measure, buying time until other sustainable forms of energy or permanent carbon dioxide removal systems can be developed. It would require large quantities of inputs in the form of management, fertilizers, pesticides and labour in order to raise the productivity significantly.

Conclusion

Biomass is an environment friendly option when assessed in terms of avoidance of GHG emissions that would have been caused by use of other alternative fossil fuels. This contribution can be exploited further by developing technologies - either *ab initio* or by adoption of other modern combustion technologies - for domestic and business/industrial applications. Liquid biofuels have opened avenues for biomass providing fuel for transport system. In this case the technology and financial viability have both been proven. Stringent environment standards and regulation will pursuade all the governments to seriously consider biomass fuels as a potential option. Only two elements are needed - efficient technology development at a rapid pace and long term national and global commitment to promote biomass for energy.

10

BIOMASS
The Sustainable Energy Source for South Asia

Introduction

South Asia is the most fragile and vulnerable region of the globe. The economic and social situation is alarming. It houses more than half of the malnutritioned and poverty ridden people. Of the 2 billion without access to commercial energy, over 800 million live in South Asia.

Various consultations organised under the World Energy Council, WEC, and the Regional Wood Energy Development Programme, RWEDP, have brought forth the need to develop appropriate policy and institutional initiatives in most of these developing countries. There is certainly a need for the policy and institutional environment in the countries to be hospitable to various efforts aimed at integrating woodfuel in wood production systems, coupled with better understanding on the use of wood energy and its impacts, now that there is a renewed interest in wood energy. RWEDP has observed that as the only existing regional level developmental assistance programme focusing on wood energy, it has played a key role in renewing such interests by developing wider awareness and deepening understanding of wood energy use and supply. As observed in various discussions capacity building activities in a number of countries in the South Asia need to be launched or strengthened to develop into concrete programmes and strategies.

In this context it is relevant to quote from the WEC Draft Rural Energy Study which presented the following scenario for South Asia in

---

Based on Paper presented at the Regional Meeting on 'Biomass as a Source of Sustainable Energy Supply' at Colombo, Sri Lanka during July 10-11, 1998 and Paper presented at the V Congress of the World Renewable Energy Network at Florence, Italy during September 21-25, 1998.

terms of bio-fuel use at Bangkok in November, 1998:

'Fuelwood and agro-residues will dominate the energy scene in the rural domestic sector. Though there is no likelihood of increased supply of fuelwood or agro-residues in near future, the total consumption of woody biomass may come down for reasons of efficiency improvement of equipment and stoves. The availability of woodfuels will be sufficient to meet the needs of the South Asia region.

Emerging advance technologies of converting and producing bio-fuels need to be promoted. The emphasis will be on final energy form rather than the primary energy form.

There will be perceptible shift towards electricity and petroleum based products in the rural energy supply in India and Pakistan. Other countries are not likely to show such a shift.

While deciding on the policy for rural energy interventions, technological advancement and financial support will be two major components.'

WEC Study on Rural Energy Supply in developing countries proved a sincere effort aimed at studying various options that could provide all the energy for rural development. Biomass was observed as one of the major actors in this region and a continued dependence on biomass well into the next century has been a major conclusion. In the changing global scenario, the primary source of input is now being relegated to back burner and the primary concern is for the final energy form. With the advancement of technologies, biomass has opened new opportunities for its conversion through modern and latest technologies into solid, gaseous and liquid forms. With the increasing commercial viability and trading of the bio-fuels the total situation is changing, though gradually. Traditionally, fuelwood was supplied or gathered free of cost. Slowly it has now been overtaken by commercial plantation. Trading of fuelwood is seen as one of the major efforts in attaining fuelwood sufficiency.

Socio-Economic Situation in South Asia

It is appropriate to consider the socio-economic conditions in South Asia before considering the energy scenario and various options of biomass energy supply.

The average rural population of all the South Asian countries was 75 per cent of national population in 1994-95. The total population of this region was about 1.2 billion. Of this population, one-half had no access to health services and almost two-thirds went without sanitation, nearly one out of four adults are illiterate and six out of every 100 children suffered from malnutrition.

South Asia covers over one-fifth of the humanity and this population of 1.2 billion is growing rapidly. More people are born in South Asia every year (27 million) than the total population of all the Scandinavian countries (Denmark, Finland, Norway and Sweden). As a matter of fact the South Asian population exceeds the total population of 50 of the smaller UN member-countries.

In the global context, the South Asia region has begun to lag behind all other regions of the world, both in its income and in human development levels. South Asia is by now (1994-95) the poorest region in the world - its per capita income of US $ 309 is much below US $ 555 for Sub-Saharan Africa and is only one-third of US $ 970 for all developing countries.

Its share in world population is 22 per cent and inhabits nearly 40 per cent of the world's absolute poor.

Adult literacy rate of 48 per cent lags behind 58 per cent rate achieved in Sub-Saharan Africa and 98 per cent in East Asia (Excluding China). South Asia also experienced an increase in its share of the world's illiterate adult population, from 32 per cent in 1960 to 46 per cent in 1993, demonstrating that adult literacy in this region increased at a much slower pace than in other regions and also that the school education was not on a desired level.

The extent of human deprivation in South Asia can be summed up as follows:

(i) Over 500 million people live in absolute poverty.
(ii) 260 million people lack access to even rudimentary health facilities.
(iii) 337 million people are without safe drinking water.
(iv) 830 million people have no access to basic sanitation.
(v) 396 million adults are unable to read and write.
(vi) 600 million people have no proper shelter.

The potential for economic growth, employment and income generation in the rural areas is centered around agricultural development, which is the source of primary food production and the base for allied activities, such as cottage and small scale industry, and off-farm village enterprises. In the face of growing population and declining land-human ratio the enhancement of agricultural productivity and other income generating activities through the introduction of modern technologies and inputs is, therefore, a major policy objective in all countries. This is accompanied by efforts to improve the quality of life in the rural areas to infrastructure development, such as access roads and communication facilities, and the provision of social amenities in the form of health care, education and water supply, among others.

Energy Sector Overview

Primary commercial energy consumption in South Asia increased at an annual rate of 6.8 per cent between 1971 and 1993. Studies project continued growth during the next several decades. ESCAP (based on IEA study) has given results for the South Asia region for the energy demand using a number of assumptions such as economic growth, energy price, technological progress, consumer behaviour and government policies on energy which need to be treated independently. IEA has given the consumption levels for the year 2000 and also projected two cases of energy demand in the year 2000 and 2010. These studies are, however, focusing on fossil sources of energy and hydro and geo- thermal power. RWEDP has indicated that the share of biomass and wood energy in the total consumption of the region will not change substantially. Overall contribution from these sources amount to about 40 per cent. Contribution of non-commercial sources of energy in the South Asia region has been significant and will continue in the similar pattern for a long time.

Very often, energy perspective does not take into account the poverty factor. Traditional energy has been central to basic nutrition and health needs of the rural poor. The low income households tend to rely on a significantly different sets of energy carriers than the rich. It is observed that the poor still depend on agro-residues, twigs and branches of the trees which are available to them free. Also, animal dung converted into cakes contributes significantly to the rural energy supply. The better off people then move on to coal, kerosene, electricity or LPG, whichever is available. Such a movement is directly related to the initial cost, cleanliness of the energy source and subsequently the efficiency of use.

The demand projections in South Asia are given in Table 10.1.

RWEDP Observation on Wood Energy in South Asia

Regional Wood Energy Development Programme (RWEDP) in Asia covers 16 major countries in South and South-East Asia. It has come out with a Regional Study titled 'Wood Energy - Today and Tomorrow'. It indicates that the developing countries in Asia consume three-quarters

TABLE 10.1: Energy Demand - South Asia - Capacity Constraints Case

LEVELS	1971	1993	2000	2010
Primary Energy (MTOE)	72	251	355	*658*
Solids	39	129	153	270
Oil	27	82	128	233
Gas	3	31	57	125
Nuclear	0	2	4	5
Hydro	3	8	12	23
Geo/Others	0	0	1	1
Final Energy	55	162	245	*461*
Solids	25	44	58	93
Oil	23	72	112	207
Gas	2	18	31	74
Electricity	5	28	44	88
Electricity Output (TWh)	76	419	650	1241
Solids	32	250	363	710
Oil	6	25	43	65
Gas	4	40	90	178
Nuclear	1	6	14	19
Hydro	33	97	139	268
Geo/Others	0	0	1	1

Source: ESCAP & International Energy Agency (1996).

of world's woodfuel used, but have only one quarter of the forest cover in the South. That sounds dramatic, but it is not .

The study also emphasises that in the 1970s and 1980s it was generally assumed that all woodfuels originated from public forest land. That strengthened the fuelwood gap theory from which it was concluded that non-sustainable yields were taken from forests to meet wood energy demand. The theory resulted in the assumption that woodfuel use was a root cause of deforestation. But, now ample evidence exists to prove that the theory is false and that, except for localised areas, woodfuel use is not a main cause of deforestation. The indication is that a major cause of deforestation is the ongoing conversion of forest land into other land uses, particularly agriculture. This is generally carried out by planned forest clearing or results from gradual processes of forest encroachment.

It has been found that about two-thirds of all woodfuels originate from non-forest land. Major implication of this observation is that woodfuel consumption is not a general or main cause of deforestation; and the woodfuel consumption will continue, whether or not there are forests. The future of Asia's tropical forests and the problems of woodfuel users are not as closely linked as is often assumed.

Economics of Fuelwood

Prof. W.S. Hulscher, CTA, RWEDP in his presentation on 'Sustainable Energy Supply Options', delivered at the Asia Energy Vision 2020 at New Delhi during November 15-17, 1996 stated as follows:

The cheapest and most reliable option for supplying renewable energy is still to grow a tree and burn it. This is actually what is commonly done in Asia, even though tree growing is seldom meant for the sole purpose of supplying energy. Unfortunately, the important roles and associated problems of wood and other biomass as a sustainable energy supply option are largely undervalued by energy planners and policy makers. A few basic data presented in the following Tables will suffice to substantiate the point.

- Table 10.2 overviews the share of wood and biomass energy consumption in South Asia.
- Table 10.3 shows the trend in biomass energy consumption (including wood).
- Table 10.4 indicates the total value of woodfuel consumption per annum estimated in terms of dollar equivalent.
- Table 10.5 summarises some figures on estimated employment per fuel type.

These data lead to the following important observations:

- Wood and other biomass supply about one-third of the national energy consumption in India, and some 37 per cent of the energy consumption in the region. Further comparisons shows that amongst the various renewable energies wood/biomass provides by far the largest share in the region.
- In all the 16 member-countries the consumption of wood and other biomass fuels is still increasing in absolute terms, even when their share in national energy consumption is decreasing.
- The total value of wood and biomass fuels amounts to more than US $ 11.5 billion in South Asia, and US $ 9 billion per annum in India (which was 30-40 per cent of export earnings

in 1991), and as much as some US $ 30 billion per annum for the 16 member-countries in the region together.

— Woodfuel use generates at least 20 times more local employment than energy from oil products (per unit of energy). Hence, national interests like balanced growth, employment and equity are linked to energy supply policies.

TABLE 10.2 : Share of Biomass and Wood Energy Consumption

Country	Year	Share of Biomass (%)	Share of Wood (%)	Share of Biomass in Domestic Sector (%)
Bangladesh	1992	72	13	89
Bhutan	1991	82	—	—
India	1992	33	—	78
Maldives	—	—	—	—
Nepal	1992-3	92	68	97
Pakistan	1993-94	47	27	83
Sri Lanka	1990	77	—	93

TABLE 10.3: Biomass Energy Consumption (PJ) in Member Countries

Country	1981	1986	1991
Bangladesh	243	162	277
Bhutan	7	9	12
India	2,165	2,441	2,824
Nepal	113	197	206
Pakistan	192	233	296
Sri Lanka	70	78	86
Maldives	—	—	—

TABLE 10.4 : Woodfuel Values in Million US$ using Average Woodfuel Prices

Country	Fuelwood	Char	Dom FW	Dom Char	Ind FW	Ind Char	Total
Bangladesh	306	—	255	—	50	—	306
Bhutan	37	3	33	—	3	3	40
India	9,080	—	8,440	—	640	—	9,080
Maldives	3	—	3	—	—	—	3
Nepal	469	—	451	—	17	—	469
Pakistan	1,318	—	1,317	—	—	—	1,318
Sri Lanka	436	5	363	—	—	—	440

TABLE 10.5 : Estimated Employment per Fuel Type

Fuel type	Tonnes of Fuel per Tera Joule (TJ)		Estimated Employment per TJ Energy Consumed in Person Days'
Kerosene	29	Kilo Litre	10
LPG	22	Tonnes	10-20
Coal	43	Tonnes	20-40
Electricity	228	MWh	80-110
Fuelwood	62	Tonnes	110-170
Charcoal	33	Tonnes	200-350

Future Role of Biomass in South Asia

Energy supply options in South Asia will include the biomass resources for domestic use in a big way. Rural households in the region are mainly dependent on biofuels. In Sri Lanka, almost 50 per cent of the households in the capital city of Colombo also use biofuels.

Dependence on fuelwood to the year 2010 is unavoidable. Introduction of higher efficiency end-use devices; and higher biomass productivities and larger areas under biomass production reassure that there will be no serious fuelwood shortage till 2010. Localised shortages may be observed, which will be mostly taken care of by fuelwood plantations on government and private lands in the non-forest areas. New initiatives in agro-forestry will promote higher productivity and inter-cropping between fuelwood species and agricultural crops. At the same time, increasing urbanisation and inter-fuel substitution will ensure containing fuelwood consumption at present level.

A small amount of demand will be met by new and renewable energy sources, such as wind power and solar energy.

Role of Fuelwood in Meeting Rural Energy Needs

Above observations have indicated that biomass resources will continue to dominate the rural energy supply in South Asia.

Observations on fuelwood supply in different countries for the year 1992-93 is given below:

Bangladesh

Bangladesh has natural forest area of 769,000 ha which is about 6 per cent of the land mass. Besides this a plantation area of 235,000 ha and other woodland of 468,000 ha also provide fuelwood. In total this is about 13.2 per cent of the land mass. The annual rate of reforestation is reported to be 11 per cent as against the deforestation rate of 3.3 per cent during 1981-90. As much as 86 per cent of all fuelwood originate from sources other than forest. In coordination with the Ministry of Agriculture, the Forest Department, under the Ministry of Environment and Forest, is also focusing on tree growing on agricultural land. Afforestation is further undertaken through the mangrove forest sources and of agro forestry projects. Agro-residues play a major role as a fuel source. The Department of Forestry has taken keen interest in developing conversion and end-use technologies such as densification technologies for fuel briquettes and efficient wood burning stoves. The technological infrastructure, however, is reported to be weak.

Among the sources of biomass fuel-cropland and livestock (leaving aquatic biomass) are privately owned. Woodfuels are obtained from three sources: grown on state owned forest land, public lands, and privately owned homestead lands. The poorer households living near state owned forests can gather woodfuel for their own consumption. Natural forests are under the management of Department of Forests. Unclassed state forests are under the ownership of the Ministry of Land and administered by the respective Hill District Councils. Village forests are privately owned and managed by the individual households.

A few constraints have been observed by the Department of Forests which include inadequate institutional capability and shortage of trained manpower, failure to design appropriate projects, insufficient project funding and inefficient forest management practices. Future fuelwood development programmes focus on reforestation, afforestation on all possible public and private lands, integrating trees with farming and traditional land use, improving the general environment for supporting agricultural and other biological production, creating employment opportunities for the landless poor, marginal savers and women and to adopt different wood conservation techniques.

Bhutan

Bhutan has natural forest area of 2,809,000 ha which is about 60 per cent of the land mass. Besides this the plantation area is reported to be 4,000 ha and other fuelwood growing areas are reported to be 355,000 ha . The annual rate of reforestation is reported at 200 ha in the year 1990. Bhutan has a major requirement of fuelwood as 97 per cent of its energy demand is met by fuelwood. It has been estimated that the fuelwood requirement is at an average of one tonne per capita per year which makes the total requirement at 1 mt per year for household application. Inspite of large forest area severe localized scarcities have been reported. The country has no comprehensive National Energy Policy and under the old Forest Act, 1969 no popular stake is incorporated in protecting and managing forest. Lack of technological infrastructure and acute shortage of qualified manpower have also been reported.

India

India has a natural forest area of 51,729,000 ha which is about 17.4 per cent of the total landmass. In addition, a plantation area of 13,320,000 ha and other woodlands of 17,689,000 ha also contribute to additional fuelwood supply. Total amount of fuelwood reported to be consumed is 94.5 mt. Share of fuelwood from the forest varies from 26-53 per cent and share of fuelwood from other sources is reported to be 47-74 per cent. The annual reforestation in 1990 was 1.009 mha per year. By 1993-94, deforestation has been arrested and the green cover has actually started increasing .

In case of India, deforestation is not due to extension of agricultural land or forest encroachment. To provide for timber needs and keeping the forest in healthy condition, India pursued a deliberate policy for import of timber from other countries; plantation of prosopis juliflora; joint forest management programme and farm forestry. The last two have led to a significant change in the Forest Policy and increased fuelwood supply in the country. The Forest Policy announced in 1988, incorporated the same. Earlier, the forest lands were reserved for timber and the village land for subsistence fuelwood needs. Since then the forests are now designated for subsistence through joint forest management; and the farmers are encouraged to grow timber as a commercial activity.

Some areas in the country face severe fuelwood problem and the situation probably has not worsened during the last few years. In other areas, local production of fuelwood has significantly helped in overcoming fuelwood shortage. Agro residues and dung have been a significant supplementary supply sources. Integrated Rural Energy Planning introduced under the Planning Commission has now been incorporated into the Ministry of Non-Conventional Energy Sources. MNES is also involved in developing the improved varieties of fuelwood species and the technologies for end use equipment. Fuelwood based gasification and dendro-thermal power generation are such significant examples. The Ministry of Environment and Forests has the responsibility for growth and maintenance of forests and has focused its attention on timber. A new Department of Wastelands Development is now focusing on growing fuelwood on lands that are not put to any other use.

Due to sustained efforts in these three ministries, improvement in end-use equipment efficiency, inter-fuel substitution and higher rate of urbanisation have resulted into lowering fuelwood demand as reported by the National Council of Applied Economic Research in 1995. It is significant to note that the fuelwood demand, in near future will be lower than the present demand inspite of the rural population going up in actual numbers.

India launched its planning process in 1951 in terms of Five Year Plans. Energy has been a major element of public support. Different forms of commercial and non-commercial sources were effectively reflected in the plans. Share of non-commercial fuels in the total primary energy supply has declined from more than 70 per cent in 1953 to less than 50 per cent in 1993. Energy planning has been the major responsibility of the Perspective Planning Division of the Planning Commission. Fuelwood has never been included in the energy planning. Integrated Energy Planning was recognised as an essential element of 'development planning' in 1960s. Energy Survey of India Committee was constituted in 1963 and the report was submitted in 1965. Its recommendations were considered in the Fourth Five Year Plan. Report of the Fuel Policy Committee was used for the Fifth Five Year Plan; Report of the Working Group on Energy Policy was used for the Sixth Five Year Plan and the recommendations of the Advisory Board on Energy were used for the Seventh Five Year Plan. The government has now set up an Energy Policy Committee which should be finalising the recommendations for the Ninth Five Year Plan to the period 1997-2002. This committee has considered biomass fuels in the overall energy supply in the country. Primary data from field surveys was collected by the National Sample Survey Organization and the National Council of Applied Economic Research.

The domestic sector energy consumption worked out by ABE, in 1985, indicated an energy consumption figure of 425 kilo calories of useful energy per person per day for cooking and space heating. This figure has been considered at 650 kilo calories of useful energy in the year 2004-05.

The Fuelwood Study Committee constituted by the Planning Commission in 1981 indicated the fuelwood consumption at 119 mtoe in 1991. It is expected that this consumption will be 103 mtoe in 2004 and 84 mtoe in the year 2009. In other words the energy supply will imply continued heavy dependence on fuelwood for energy during the next 10 years. In the case of both, the crop residues and the dung, the competing demands would also increase simultaneously. It is difficult to visualise any incremental availability of either dung or crop residues without its further adverse impact on soil fertility and livestock productivity.

Maldives

Maldives is a cluster of small coral islands. The total forest land and wood land is reported to be about 1,000 ha which is about 3.3 per cent of the total land mass. It should also be noted that the total arable land was only 10 per cent of the total land area in the year 1993. As the economy of the country is based on tourism, most of the commercial activities are dependent on the imported fossil fuel. The natives have, however, to depend on fuelwood as the major source of energy supply. The Commonwealth Secretariat had initiated action for survey of fuelwood resources and their identification in 1987. Also, under their support, designs of improved cookstoves were introduced. However, due to very little land mass for fuelwood cultivation, sufficient quantity of fuelwood is not available on each of the islands. Although the country has a law which makes it mandatory for anyone cutting one tree to plant two, but that is never enforced.

Nepal

Nepal has natural forest area of 5,023,000 ha which is 37 per cent of the total land mass. In addition, it has a plantation area of 56,000 ha and other woodlands of 672,000 ha. Total quantity of fuelwood consumed is reported to be 11.3 mt per year and the share of the forest wood is 66 per cent, whereas share of wood from other sources was reported to be 34 per cent. The energy balance for the year 1995-96 has indicated that about 4.3 mt of fuelwood is supplied by over-exploitation of forests. Estimated area of forests and shrub land, currently assessed, is around 3 million ha. For the country as a whole 82 per cent of the total fuelwood requirement is supplied from off-farm and remaining 18 per cent from the farmland. In the year 1995-96, sustainable fuelwood supply from accessible forests, shrub and grassland is estimated to be about 5 mt; 1.1 mt from farm land and 0.7 mt from non-cultivable inclusions. These supply figures indicate that around 65 per cent of the total fuelwood requirement of the country is sustainable. Non-commercial energy accounts for 92 per cent energy consumption in 1992-93. Fuelwood is important in Nepal for cooking as well as for space heating. The government's policy on forest management aiming to turn the forest back to village and user groups has shown successful results. New experiments have been started with regard to system of private leasehold forests.

Fuelwood requirement of urban areas has been met by forests situated upto a distance of 400 kms. Transport of fuelwood requires use of imported commercial fuel. Such transported fuelwood has more than 60 per cent moisture content based on oven dry weight . Therefore, in the energy content scenario a lot of foreign currency has been unnecessarily drained in fuelwood supply to the urban areas located long distances from the source. In most of the rural hilly areas fuelwood supply is non-monetized and it has been readily available in unrestricted quantities in close proximity to the consumption point.

The fuelwood development is a part of the government's policy. The last Five Year Plan (1992-97) stated that one of the main objectives of forestry sector development was 'stability of supply of timber, fuelwood, fodder and other forestry products necessary for common people in their daily life'. The Forestry Sector Master Plan prepared in 1988 realised the importance of Wood Energy Plan. The document clearly identified the imbalance in the fuelwood supply and demand and suggested methods of augmenting fuelwood supplies, with an emphasis on community forestry development. Nepal has also developed a good wealth of data on fuelwood energy and rural energy supply and demand. Studies in the five different regions in Nepal. An overview of fuelwood supply and demand, have been completed. An Energy Pricing Study has been conducted for flow of fuelwood in Kathmandu valley. A Perspective Plan has been completed in 1995 incorporating fuelwood and biomass energy as one of the five components. Special attention has been paid to fuelwood energy planning realising that a reduction in fuelwood energy consumption is not foreseeable in the near future. The Water and Energy Commission Secretariat has been the main government department for conducting such policy guidelines.

Traditionally, fuelwood trade was carried out in many urban areas by head-loaders and is in practice even today. 'Fuelwood Cooperatives' were marketing fuelwood through its numerous depots in different urban centres until recently. Department of Forests now controls the fuelwood cooperative. All urban fuelwood marketing and supply system in Nepal has been almost completely privatised. To some extent fuelwood supply in the fuelwood scarce rural areas is also managed by user groups under the community forest and leasehold forest programme. Under this programme, scattered patches of natural forests are being handed over to local communities for overall management by user groups with nominal technical support.

Pakistan

Pakistan has natural forest area of 1,855,000 ha which is 2.4 per cent of the total land. In addition, a plantation of 168,000 ha and other woodlands of 1,105,000 ha are also contributing to fuelwood supply. The annual rate of deforestation has been reported to be 2.9 per cent during 1981-90 and a corresponding annual rate of reforestation is reported to be 4,200 ha per year. In view of the limited forest area, the country has to focus mainly on agro-forestry, which is successful now. In late 70s, poplar species were imported from Italy and is now widespread in Pakistan, sustaining substantially the local wood industry and fuelwood supply. Almost 98 per cent of all fuelwood comes from agricultural land, and agro forestry has greatly improved fuelwood supply. About 1.5 per cent of the agricultural land has been put under tree cover and a suggestion is made to raise it to about 5 per cent without reduction in agricultural production. The forestry sector programme implementation takes place at the provincial level. The Ministry of Forests and the Ministry of Environment have important role to play. World Bank/ ESMAP Study on Household Energy Survey (SHES), 1991-93, was the most comprehensive study on rural energy supply. Pakistan Forestry Institute has a satisfactory technical manpower base.

Sri Lanka

Sri Lanka has natural forest area of 1,746,000 ha which is about 27 per cent of the land mass. Besides this, a plantation area of 139,000 ha and other wood lands of 2,113,000 ha also contributed to additional fuelwood supply. The biomass fuel consumption contributes to 53 per cent of total energy consumption in 1992. The annual rate of deforestation is reported to be 1.3 per cent during 1981-1990. The country has national forestry programme which also addresses bio-energy as an important energy source for Sri Lanka. Biomass, including fuelwood and charcoal, figure in the national energy balance of Sri Lanka for the last 10 years. (This has been confirmed by the Ministry of Irrigation, Power and Energy). An important programme of improved wood stoves was introduced which was supervised and implemented by the Electricity Board. The stove programme for both the rural and the urban areas is now promoted with the help of private entrepreneurs.

Biofuel Programme other than Fuelwood

Except for India no other country in the region has a large programme of biofuel use. India is the only country that launched major programmes in biofuel.

Nepal has undertaken biogas programme on a small scale. Nepal, Sri Lanka and Bangladesh have large scale demonstration programme of improved high efficiency cookstoves for burning fuelwood.

It is relevant to give examples of programme in India alongwith the number of installations.

Biofuel Programme in India

National Programme on Biogas

National Project on Biogas Development, NPBD, initiated in 1981-82, aims at promotion of family type biogas plants for providing clean and cheap source of energy in rural areas, producing enriched organic manure for use in agricultural land, improving sanitation and hygiene, and removing drudgery of women. Against a potential of setting up 12 million biogas plants based on cattle dung, 2.7 million biogas plants have been set up (upto March 31, 1998).

The programme for community, institutional and night soil based biogas plants was initiated in the year 1982-83, with the objective of setting up large biogas plants in the villages and at the institutions having large quantities of cattle dung and the need to utilise biogas for engine operation for mechanical power and electrical power generation.

Various R&D programmes have been initiated to study the micro-biological processes to increase biogas yield in the field. Efforts are also being made to reduce the cost of biogas plants by using alternative building materials and construction methodology. The Annual Report of the Ministry of Non Conventional Energy Sources, Government of India for the year 1997-98, has reported that 2.7 million family biogas plants and 2,300 community biogas plants in the country generate biogas equivalent to saving of 8.6 million tonnes of fuelwood per year. Besides, these plants are generating about 41 million tonnes of organic manure per year for supplementing chemical fertilisers and improving fertility.

National Programme on Improved Cookstoves

National Programme on Improved Cookstoves started in 1985-86, aims at achieving fuelwood conservation and thereby checking deforestation and environmental degradation; reducing smoke from kitchen; removing drudgery of women and girl children and generating employment in the rural areas. The programme has an estimated potential to cover 120 million rural and semi-urban households. These improved cookstoves are expected to save about 25-30 per cent of the fuelwood consumed in a normal cookstove.

Annual Report of the Ministry of Non Conventional Energy Sources, Government of India, for the year 1997-98 has reported that field studies have shown that the improved cookstove, on an average, saves about 400 kgm of fuelwood equivalent per year. Thus 28.3 million improved cookstoves set up so far in the country save about 11 million tonnes of fuelwood every year.

Biomass Use Programme

The routes followed in India for efficient use of biomass for industrial application have been - gasification and briquetting. Biomass gasifier converts fuelwood and agro residue into producer gas through thermo-chemical process, which can be used for thermal and mechanical or electrical applications. Fuelwood based biomass gasifier systems of upto 500 kw capacity have been developed indigenously and are manufactured in the country. Similarly the technology for making biomass briquettes from agricultural residues and forest litter, at both household and industrial levels, have been developed.

Recent Energy Policy Initiatives

The general thrust of the macro-economic reforms, mainly in India and Pakistan, is to reduce the extent of public investment and intervention in general by encouraging participation of the private sector in national development.

Arising from the above, a key policy shift has taken place within the energy sector is to explore the possibilities of privatising the energy supply industry and/or inviting the participation of the private sector in energy development. Different approaches have been adopted by different countries in pursuit of such a policy shift. Participation of the private sector in developing new power generation capacity in India and Pakistan are examples of these developments.

As an extension of the policy concerning energy development, the involvement of the private sector in energy supplies to the rural areas has become a matter of interest in some countries. Although, the concept, and the mechanism to operationalise it, are yet to crystallise, the number of private energy enterprises has been growing in some countries (India, in particular). A related development is the policy to promote self-generation and co-generation in agro-based industries (sugar mills, palm oil factories, rice mills), and to assess the potential of excess power generation that could be sold to the grid. Such arrangements have been in existence in Malaysia, Philippines and Fiji, although the electricity generated by the private sector is mainly for its own consumption and not sold to the grid. By contrast, some state power utilities in India have introduced the concept of 'energy banking' schemes to promote the development of wind energy. Biomass based cogeneration programme of 25-30 MW has been attracting interest of a number of small and medium industries from other countries. The level of financial implication is low and they can easily promote a large number of such units. This route opens opportunity for private sector cooperation outside government's involvement.

Greater attention is also being paid to participation of rural energy users in co-financing rural energy projects, in particular, projects for decentralised energy systems and fuelwood plantations.

Interventions and Institutions

Interventions in the wood energy sub-sector largely depends on government actions like public resource management, pricing, taxes, subsidies, economic incentives etc. Such interventions can only be prepared and implemented when the country avails of adequate institutions and sufficient number of qualified experts. Furthermore, interventions with regard to wood energy are to be supported by the establishment of effective public-private partnership for introduction of new supplies and managing existing ones; and for private sector organisations for production, conversion and utilisation technologies.

Wood energy, that has direct relevance to rural areas, has still not found appropriate place in National Energy Policy. More so fuelwood does not find an appropriate place either amongst the ministries dealing with energy or with agriculture and forestry.

Next comes the development of skills for units dealing with developing, planning and managing wood and biomass energy. In order to gain momentum a critical mass of educated and well trained people will be necessary. Fuelwood development is not treated separately and is covered under forestry sector in each country. They have good forests, but often they do not focus on fuelwood planning and supply.

Further, important aspect of institutional development relate to the role of state and local government authorities, NGOs, community participation and private sector participation.

RWEDP has observed that the starting point will be fuelwood energy planning; and integration of such plans with development plans of agriculture, rural development, forestry and renewable energy plans.

RWEDP as an institution has developed a strategy for building skills in its member-countries for data collection, analysis and planning for fuelwood production, utilisation and meeting demand. The coordinating role of RWEDP has made it possible to conduct study courses, exposure visits and training workshops in various countries in this region. A number of activities have been identified, where regional capabilities have improved skills through technical cooperation or collaborative programmes. The effectiveness of such a programme lies with experience earned in different activities by the member-countries and overall improvement in skills.

Impact of Technology on Biomass Use

Following approaches in biomass conversion technologies have been advised:

(i) The 'traditional domestic' use in the developing countries for household cooking, lighting and space heating. In this role , the efficiency of conversion of the biomass to useful energy has been raised from 5 per cent to 20 per cent.

(ii) The 'traditional industrial' use of biomass for the processing of tobacco, tea, pig iron, bricks and tiles sector, where the biomass feedstock is often regarded as a free energy source. There is generally little incentive to use the biomass efficiently and so the conversion of the feedstock to useful energy commonly occurs at an efficiency of 15-30 per cent.

(iii) The 'modern industrial' activities relate to thermal conversion technologies. Expected conversion efficiencies are between 30 and 55 per cent.

(iv) 'Newer chemical conversion technologies (fuel cell)' are capable of bypassing the entropy dictated Carnot Cycle limit which describes the maximum theoretical conversion efficiencies of thermal units.

(v) 'Biological conversion techniques' including anaerobic digestion of biogas production and fermentation of alcohol

India has undertaken major research and development programmes in the emerging technology; and large scale demonstration programmes in traditional domestic and industrial applications. Besides India other countries have only undertaken activities related to traditional domestic and industrial use.

Results of R&D efforts in India have helped in stabilising the efficiency improvements and thereby conservation of biofuels in different devices. Significant to mention is the improvement in efficiency of cookstoves, higher gas production from biogas plants (even at low temperatures) and consistent performance of gasifier and briquetting machines. These are examples of the level of engineering perfection achieved in the country, and reflects an important reason for wider use of such devices in India.

Rural Energy Supply in South Asia—Future Directions

The future rural energy demand and supply will have to take into consideration the following realities:

(i) The lack of modern forms of energy supply limits the opportunities to achieve major structural changes in the rural economies that could lead to enhancement of the income generating activities and poverty alleviation.

(ii) In the absence of appropriate energy alternative the rural people are deprived of basic social amenities, such as health services, sanitation, water supply and education.

(iii) Time spent on collection and using traditional fuels represents a significant loss of human availability for self-development and productive economic activity. Supply will have to be smoothened so that the rural population spends its time in a productive fashion.

(iv) The chances of eliminating dependence of the rural poor on traditional energy sources by the year 2010 are slim.

(v) Fuelwood and agro-residues will dominate the energy scene in the rural sector. Though there is no likelihood of increased supply of fuelwood or agro-residues in near future, the total consumption of woody biomass may come down for reasons of efficiency improvement of equipment and stoves.

(vi) Renewable sources of energy are expected to contribute a greater percentage of rural energy supply. However, devices based on woody biomass and biogas plants have a perceptible future as they will be of low cost.

(vii) There will be a perceptible shift towards electricity and petroleum based products in the rural energy supply in India and Pakistan, other countries may or may not indicate a similar change.

(viii) Regional cooperation and appropriate financial support can lead to higher capacities of hydro electricity installation in Nepal and Bhutan which may supply electricity to the rural areas in Bangladesh and North Eastern India also, (besides in the country of origin)

(ix) Similar regional cooperation may also lead to India and Pakistan getting larger quantities of natural gas supplies from the Middle East and Central Asia. This may result in natural gas availability also in the rural areas, both for industrial and domestic applications.

(x) While deciding on the policy for rural energy intervention - technological advancement and financial support will be two major components.

Certain Issues for Consideration

Certain issues need to be carefully examined and addressed to develop biomass programme in South Asia. These can be categorised in the following sub-groups:

— Policy and Planning
— Strategy
— Research and Development
— Environmental Aspects
— Regional Cooperation

These categories of issues can be detailed as follows:

Policy and Planning

(i) The Energy policy of each country should take into consideration, the biofuel consumption, in quantum terms, especially the woodfuel consumption, in the country and its increasing role. Fuelwood as a major energy source in the rural domestic sector is well established. The increasing role of woodfuels through high efficiency conversion also make them attractive for the industry. Therefore, the role of biofuels in household and industry sectors in rural and urban areas should be appropriately reflected in the National Energy Plan.

(ii) Decentralised Local Level Planning, based on natural resources, should consider fuelwood and other biomass fuels as main element of Village Development Plan. Bio-energy development should have integration with agriculture, forestry, horticulture, animal husbandry, irrigation and rural development. Local communities should be involved in planning and implementation.

(iii) National Wood Energy Data Bases should be set up in each country. More key micro-level data on wood energy consumption pattern and wood energy supply should be collected to support Wood Energy Policies. Access to such information should be provided to various private and public sector agencies.

(iv) Land Use and Cover Policy should be formulated with incentive mechanisms to provide intervention of woody perennials in the system where they are needed most.

(v) Prevailing rules and regulations which hamper wood energy development should be reviewed. Wherever necessary, the land ownership and holding tree tenure, tree planting and harvesting in private and community lands, transportation and trade of wood and related products produced by the private sector or local communities should be modified to support higher biomass growth.

(vi) Support price mechanism for woodfuels, on the lines of agricultural crops, should be introduced to organise marketing and assured returns on energy plantations.

(vii) Policy Research and Planning should be encouraged to minimise regional imbalances and local shortages.

Strategy

(i) The local level governance, which in case of India and Nepal is called the village level panchayats, should be effectively utilised for higher biomass production and utilisation. In case of India, the benefit of the 73rd and 74th Constitution Amendments should be taken to impress on the third level governance - district, block and village panchayats - to be used appropriately and effectively as a delivery mechanism for providing income generation opportunities to the rural youth in villages through appropriate use of locally available biofuels.

(ii) A package of practices for growth of food crops and energy crops on the same land should be developed to ensure highest biomass growth and should be disseminated through extension agencies and community based organisations.

(iii) Agricultural Research System (and not Forestry System) should develop a programme for higher biomass (fuelwood) production and its linkage to income generation activities. If necessary it may collaborate with other major research institutes and laboratories and seek collaboration with International Agencies for funding and technical cooperation.

Research and Development

(i) Selection of fast growing species for wood energy crops, identification of appropriate provenance to match specific conditions, and improvement of the survival and growth rates of trees at degraded sites and wastelands should be supported by further R&D.

(ii) R&D efforts in the following biomass production activities should be focused:

— Collection, evaluation and improvement of woody perennials for fast growth and high production efficiency.
— Blending species of herbaceous and woody perennials in agroforestry systems to capitalize on complementaries.
— Research in plant geometry and canopy management of woody perennials for improving their production efficiency.
— Research in below ground interactions, fine root dynamics and nutrient turnover under different cropping systems to harness their potentiality for high nutrient efficiency, resource saving and preserving/improving environmental quality based on exploitation of synergies.
— Research in the potential of tree species as biopumps in water logged areas and bioremediation of polluted sites.
— Wood preservation, seasoning and post harvest technology for value addition.
— Role of leguminous trees in soil improvement and production sustainability.
— Blending of annuals, biennials and perennials in a holistic system in a systems approach.

(iii) R&D for upgrading conversion of fuels from forest residues and other biomass should be promoted. R&D efforts in the following bio conversion technologies should be promoted:

— Development of effective binding agents for briquettting.
— Development of conversion devices resulting in lower CO emissions, low smoke and reduced health hazards.
— Development of low cost technologies for extraction of alcohols and plant easter oils from agricultural waste materials.

(iv) Quality seeds should be grown, collected and properly stored. Dormancy Test on seeds should be conducted.
(v) Integrated Plant Nutrient Management practices should be promoted for effective higher biomass production.
(vi) Biofertilizers, mainly the mycorrhizial technology, should be promoted for low cost reclamation of wastelands and promoting higher fuelwood plantations.

Environmental Aspects

(i) Social, economic and environmental roles of woodfuels

produced in both the forest and non-forest areas should be recognised; and woodfuels should be treated as an important sub-sector which needs to be developed.

(ii) Cost effectiveness of wood energy development projects should also incorporate environmental cost of the same.

(iii) Long term carbon dioxide sequestration potential of sustained growth of biomass has the environmental advantage. Additional investments can be fully justified.

Regional Cooperation

Some of the issues for Regional Cooperation can be defined as follows:

(i) Development of Regional Research and Development Programme on identification and production of short rotation fast growing fuelwood tree species.

(ii) Development of a Regional Database on biomass demand and availability in the region.

(iii) Energy - Industry Collaboration

(iv) Sharing of Technological Resources

(v) Developing Regional Funding Mechanism

(vi) Sharing Investments in Energy Conversion Systems

(vi) Development of Human Capital Resource

(vii) Strengthening Regional Institutional Mechanism for technical cooperation and the skills building.

Conclusion

South Asia needs energy supply additions at a fast rate. However, lack of financial and technological capabilities hamper its progress. The poor rural population has depended on traditional biomass for energy supply for living at the sustenance level. This situation will not change dramatically. Biomass, will meet the energy needs well into the next century.

Biofuels will continue to meet energy needs in South Asia. New Technology development will support conversion of solid fuel into liquid and gaseous forms. Regional cooperation will be crucial for developing human and financial capital for the region.

11

BIOGAS AND RECENT EXPERIENCE WITH ANAEROBIC DIGESTERS IN INDIA

Introduction

Energy supply in South Asia and Sub-Saharan Africa will largely depend on non-commercial fuels. Rural areas in China will also be largely dependent on non-commercial fuels. Switching over to commercial fuels will always be a desired goal. But, due to economic compulsions the non-commercial renewable sources will have a significant role. Possibility of anaerobic conversion of residues and wastes into useful form of energy sources will always remain a high priority. Certain countries like Nepal and Myanmar will depend on non-commercial sources to meet upto 90 per cent of their total energy demand. Countries with large animal populations have focused attention on biogas production. India and China has a large number of biogas plants - and thereby they have generated information based on their experience. Neighbouring countries are learning from them.

Concept of Family Biogas Plant

Biogas production is a major revolution in the developing countries for conversion of cow-dung into useful gas and at the same time providing much needed micro-nutrients to the agricultural lands. Methane gas is produced naturally by the crackdown of organic matter. This gas has been recorded as 'marsh gas', 'dancing flame', and 'fuel fire', etc.

The first report of the production of gas from a carefully designed septic tank in 1895 came from England; and by 1920 devices were made producing methane gas. In China, Luo Quorni constructed the

---

Subscribed to Special Publication of World Energy Council, London in July, 1998.

earliest digester in 1920. During the Second World War, the German developed a prototype of biogas plant for overcoming the shortage of fuel. After this war, biogas generation was adopted in France and USA. In India, though the first anaerobic digester was set up in 1900, interest was renewed in 1937 with the commissioning of a Sewage Purification Station at Dadar, Bombay. Later, scientists of the then Imperial Agricultural Research Institute, the present IARI, Pusa, New Delhi came up with a patent on cattle dung digester in 1946. In 1951, Mr. Jashbhai Patel patented a practical plant and named it 'Gram Lakshmi'. This design was adopted by IARI; the Planning, Research and Action Institute, Lucknow; and the Government of Maharashtra. All these plants were designed and operated on the basis of animal dung output of 5 animals. This concept is still prevalent in India, and the small size family type biogas plant is based on animal dung from 5 animals and gives an output of 2 cum gas per day.

Biogas plant operation depends on anaerobic digestion which requires a good substrate. The organic residues suitable for methane generation fall under three groups, viz, animal, crop and agro-industrial wastes. Typically the animal wastes serve as the major feed-stock as they are potential source of carbon and nitrogen. Though various industrial wastes can also serve as substrates for anaerobic digestion, their digestibility can be improved by mixing them with animal wastes in appropriate proportions to maintain the C:N ratio at 30:1.

India has a potential of 12 million digesters. So far 2.7 million digesters have been installed. Also, 2,300 community type large biogas plants have been installed so far in the country. These are estimated to generate fuel gas equivalent to saving of about 8.6 million tonnes of fuelwood per year. Besides, these plants are generating about 410 million tonnes of enriched organic manure per annum for supplementing chemical fertilizers and in improving fertility. Apart from this about 7,000 cubic metres of biogas is estimated to be generated per day from community toilet complexes and utilised for operating dual fuel engines for drinking water supply in slum colonies (up to March 31, 1998).

China has over 5 million biogas digesters, these categories are both in the domestic low cost type of biogas digesters. China is also reported to have 10,000 industrial digesters and over 800 biogas electric power stations.

Biogas Use Potential and Problems in Developing Countries

Family size biogas plants are extensively used in the poor regions of

Asia and Africa. Their experiences are valuable indicators for future R&D.

The main technical problem faced by biogas system in China and India, which have the most extensive experience with biogas technology in Asia, have been the poor sealing properties of digesters, low biogas production, difficult discharges of fermented materials, imperfect pipeline systems, and problems with pre- and post-treatment systems. The Indian floating drum design has generally proven more reliable than the Chinese fixed dome design. Over the years, the costs of both family and community size plants have come down and reliability factors have improved considerably. The principal barriers to biogas technology are the difficulties in obtaining social acceptance and consequent uncertainties in the regular supply of biomass feedstock.

On the other hand, small scale biogas systems in the rural household have not been successful in Africa due to two factors: (i) the unacceptable complex management and maintenance back-up they require; and (ii) the absence of operational experience at the household level. On the other hand, small scale biogas projects targeting rural institutional applications have demonstrated encouraging results. Valuable field evidence indicates that combined septic tank/biogas units run by institutions such as rural hospitals and schools represent attractive option for introducing biomass technology in rural areas of Africa.

Research and Development on Biogas

A review of worldwide activities relating to biogas technologies has identified the following thrust areas:

- Studies in the field of micro-biology, bio-chemistry and engineering for increasing the biogas yield by at least a factor of two, especially at low and high temperature,
- Development of cost effective designs of biogas plants,
- Development of design and methodologies for utilisation of other biomass and agro-residues for biogas production,
- Cost reduction of biogas plants by use of alternative building materials and construction methodology, and
- Diversified use of digested slurry for value added products.

Experiences in Different Developing Countries

Study on Fluidised Bed Anaerobic Digester, AIT, Bangkok

Whereas animal dung is traditionally used for anaerobic digestion, other systems have also been tried. The anaerobic contact process was applied to waste of several processes. Lack of fundamental knowledge of the process and its limitations have probably limited more widespread usage of this process. In a study conducted by Asian Institute of Technology, Bangkok, under the guidance of Prof. S.C. Bhattacharya, on fluidized bed anaerobic digestion process the anaerobic filter for the treatment of concentrated soluble organic waste waters was observed to treat dilute waste in a short period of time in ambient temperatures.

This concept has been applied to anaerobic system found to be highly efficient at short hydraulic times. Pilot plants containing large scale fluidized bed bioreactors as advance treatment process for nitrification and de-nitrification of industrial waste water has been successfully employed. A similar process has been employed to remove nearly 80 per cent BOD at high organic loading rates with a simultaneous generation of methane from stabilised high strength wastes including from dairy, chemical, food processing, soft drink bottling and heat transfer liquids.

The laboratory scale model was used to study the process of fluidized bed anaerobic digestion is suitable for waste stabilisation.

Chicken Dung Biogas Power Generating System in Malaysia

Utilisation of biogas from animal waste in Malaysia was considered in the early 80s. However, it was felt not attractive at that time, mainly because there were no large scale livestock industry and that farm animals were normally scattered so that collection of waste was difficult. Subsequently a study on pig-waste biogas was conducted on a unit being fed by 500 to 600 pigs at its research station. The pig waste based biogas system provided enough data for utilisation of other wastes. A study was conducted in Kualalumpur on a farm consisting of 300,000 heads of layer. This system is expected to supply 60750 MJ. It can generate energy equivalent to 70 kw generator operation for the day.

Biogas Production from Lignocellulosic Material in Nigeria

This experiment was carried out to study the effect of temperature on biogas production. The cow dung used in this investigation was obtained from a local cattle market and the reactors were also fabricated locally. It was observed that the gas production continued to increase from the first to fourth week after which the rate decreased for the sixth week. A slight increase was observed in the 7th week and subsequently there was a decline. Results also indicated that for digesters maintained at 60°F produce more gas than those maintained at 40°F. At 27°F only small quantity of gas was produced. The study also indicated that methane producing bacteria was very sensitive to sudden changes. Therefore, any drastic change in the temperature should be carefully avoided.

This kind of experiment in Sub Saharan Africa is not very common.

Technical Evaluation of Rural Biogas Installations in Tunisia

The research work carried out in Tunisia was aimed at evaluation of the research results for effective dissemination in the rural areas. Tunisia uses biogas installations of the fixed dome type like that of the Chinese design. As the number of biogas plants are very small sufficient information and data has not been generated on the same. The study concluded that lack of knowledge of systems and technical capability for maintenance and repairs has been a major constraint in larger dissemination of biogas plants.

These are a few examples of experiments carried out on the biogas plants in different parts of the world, mainly in the developing countries. It gives an idea about action being initiated to improve and make biogas systems foolproof.

Research Work in India

Research or development on biogas production systems is an ongoing programme in India. Systematic project implementation takes place every year.

During the year 1997, following projects have been completed in India.

(i) Under the project on 'Multistage Digester System Development for Conversion of Municipal Solid Waste', the Nodal Research Centre, Calcutta has designed a four chamber anaerobic digester. Each chamber of the digester has stirrer. The digester was fed with a slurry of crushed municipal solid waste. A filter bed was also designed for drying of treated slurry. An average gas production of 80 litres per kilogram of waste was recorded.

(ii) Under a separate project entitled, 'Ferro-cement biogas digester (domestic scale), Centre for Scientific Research', Auroville, Tamil Nadu has completed field testing of low-cost ferro-cement Deenbandhu model biogas plants and recommended its promotion through entrepreneurship.

(iii) A laboratory scale UASB reacter has been set up to study parameters, such as volatile acids, chemical and biological oxygen demand, and gas production under the project entitled' 'Factors affecting granulation of methane producing bacteria in UASB reactors' at Regional Research Laboratory, Council of Scientific and Industrial Research, Thiruvananthapuram.

(iv) Research work is in progress on 'High temperature anaerobic digestion of thermophilic anaerobes' at Tamil Nadu Agricultural University, Coimbatore.

(v) Under another project on 'Dissemination of leafy biomass based biogas plants' is going on at IISc. Bangalore, two biogas plants each made of burnt bricks and stabilized mud blocks have been set up for field trials. Also biomethanation potential of seven biomass substrates has been assessed.

(vi) Studies on 'Methanogenesis under low temperature conditions for optimum production of biogas from agricultural residues, cattle dung etc' have been in progress under another project. Samples collected from biogas plants functioning in hilly areas of Uttar Pradesh for isolation, screening and selection of methanogenic consortia at low temperatures have been standardised. Further studies on optimisation of process parameters for efficient generation of biogas are in progress.

Conclusion

Biogas offers an attractive fuel to the user. However, the generation of biogas is dependent on local environmental conditions and the kind of input material. Its wider use demands intensive R&D for making systems user-friendly.

REFERENCE

Annual Report, Ministry of Non-Conventional Energy Sources 1997-98, Government of India.

NOTE

This Paper was written jointly with Prof A.K.Ghose.

12

PEOPLE'S PARTICIPATION IN INTEGRATED RURAL ENERGY PROGRAMME IN INDIA
A Case Study

Introduction

Rural energy is a comprehensive term which reflects a combination of different sources of energy that can be used for meeting all the needs in the social and income generation sectors in the rural areas of India. Lack of access to commercial energy for basic needs has made the rural people dependent on fuelwood, cowdung and agricultural waste. Energy for productive activities is also scarce in rural areas for the same reasons.

Integrated Rural Energy Planning (IREP) Perspective

The Planning Commission, Government of India initiated a comprehensive integrated programme including all sources of energy, on a pilot basis at the micro-level of a block (the smallest development unit in the country). As a part of these integrated rural energy plans, most cost-effective mix of different energy sources has been worked out for meeting various needs of subsistence and production, in keeping with the specific situations and resource endowment of the macro region. The specific components of IREP included the following:

— Development of institutional mechanism in Federal States/Union

---

Paper presented at the National Workshop on 'Integrated Rural Energy Programme' organised by the Non Conventional Energy Development Corporation of Andhra Pradesh at Hyderabad on April 25-26, 1992.

Territories;
- Training;
- Project preparation;
- Project implementation;
- Provision of financial incentives; and
- Monitoring.

The Integrated Rural Energy Planning Programme was initially introduced in 250 blocks (out of a total of over 5,000 blocks in the country).

As a result of the success achieved during the Seventh Five Year Plan (1986-90), IREP was expanded under the Eighth Five Year Plan (1992-97). Specific components include the following:

(i) Extension for IREP Programme to cover at least one block in each district of the country—proposed to extend at a rate of 100 blocks per year during the Eighth Plan;
(ii) Provision of technical and financial support for IREP block projects including staff, planning and implementation and development of planning capabilities; and
(iii) Implementation.

The emphasis shall, however, be placed on the following:

- Meeting basic energy needs for cooking, heating and lighting, especially for economically weaker sections;
- Increasing employment and income especially of small farms through increased agricultural production, by planned inputs of energy;
- Involvement of voluntary organisations, panchayats and other local and planned people's organisations in implementation of IREP Projects;
- Increasing employment by energising village and cottage industries under IREP Projects;

Thus, it will be seen that the Integrated Rural Energy Planning Programme initiated in the Seventh Five Year Plan have now developed into a full fledged Integrated Rural Energy Programme.

This integration is of two kinds as follows:

(i) Integration of different sources of energy - commercial, conventional and non-conventional - for all rural needs;

(ii) Integration of income generation activities with the availability of energy for overall area development.

Thus, this approach is likely to lead to a wider integration of development and energy in the rural sector of our country. Integrated Rural Energy Programme is still in the age of evolution, but the approach fulfill a very specific need of the rural areas.

People's Participation—An Important Component of IREP

The components of the IREP and the areas of emphasis, as described above, clearly bring out the point that this programme has two important aspects for which people's participation is of utmost importance:

(i) It is aimed at 80 per cent of the population;

(ii) Voluntary agencies, panchayats and local bodies are to be effectively utilised for implementation.

It will be noticed that such a shift in implementation of IREP basically suggest that the responsibility for implementation will be shifted from the State Government Nodal Agencies to the Voluntary Agencies and the democratically elected local units of village panchayats.

This approach can be looked at as an integration of energy into the people's movement and into the social fabric of rural India.

With a view to study the likely impact of such an approach and also the interest of the common man in income generation programmes by effective use of different sources of energy, it was considered necessary by the Indian Association for the Advancement of Science, IAAS, to conduct studies in some of the blocks which were directly covered under the I.R.E.P Programme. One such block in the hilly region of Uttar Pradesh was studied thoroughly with a view to identify people's interest in developing income generation and employment generation opportunities by effective use of available energy sources. The Block is Betalghat in district Naini Tal.

Case Study of An I.R.E.P. Block—Betalghat

A survey was conducted by the Indian Association for the Advancement of Science in block Betalghat, District Naini Tal in the Kumaon hills of Uttar Pradesh during the middle of 1991.

Location

This location was chosen as a 100 kilowatt micro hydel scheme was launched by the Non-Conventional Energy Development Agency, Uttar Pradesh, using the experimental low head cross flow turbine developed by the Department of Science and Technology, Government of India. It was observed that almost 70 kilowatt of this perennial source was available for use for income generation activities.

Objective

The objective of the survey was to study the feasibility of use of off-peak time power from micro-hydel power plant in hilly area for income generation and employment generation.

Survey Methodology and Coverage

The survey methodology was decided to include a large area in the block so that the people who may be benefited from the proposed scheme or project were represented. Thus, a sample of ten per cent of total villages in the block was selected, covering sixteen villages. The sample of household to be covered varied from 50 to 100 per cent depending on the population of the village. This high percentage was decided as the population of the hilly villages was very low. Betalghat block has 160 villages spread over 21,140 ha out of which 9,935 ha was under cultivation, and total population was 42,400.

The survey methodology aimed at identifying use of different sources of energy for domestic sector, agriculture sector and other income generation (small, cottage industries and artisanal) activities. People's reaction and the strategy for promotion of people's participation was specifically covered.

Thus, the methodology was so decided that survey provided very important input for initiating programme for people's participation in implementation of I.R.E.P.

Observations

Survey observations can be summarised as follows:

Population Distribution

Population Distribution in surveyed villages indicated 28 per cent over 35 years of age and 72 per cent below 35 years of age. Thus, the block has a strong population percentage which is in the productive age-group.

Literacy Rate

Literacy rate in the age-groups upto 35 years indicated almost total literacy for male population and upto 93 per cent for female population. Corresponding figures for age group over 35 years indicated literacy in male at 88 per cent and literacy in female at 36 per cent.

Main Source of Family Income

The main source of family income was agriculture. Even in such cases where the primary source of income was business/service/craft, agriculture remains an important secondary source of income. The distribution of families with agriculture as a primary source of income was 81 per cent and that of the secondary source of income as 18 per cent. Family income was mostly upto Rs.3,000 per month with even distribution in different blocks of upto Rs. 1,000 per month, Rs.1,000-2,000; and Rs.2,000 – 3,000.

Cereal Production

Cereal crops grown were mainly wheat, rice, maize, pulses and soyabean and bathuwa. Low productivity and low inputs were observed.

Horticultural Crops

New and hybrid varieties developed by different research laboratories have been suitably extended and vegetables like pea, capsicum, tomato, cauliflower and brinjal are being grown. These varieties are also suitable for growing during the off-season.

Energy Consumption

Energy consumption in the domestic sector makes an interesting reading with fuelwood meeting 97.1 per cent of the needs for cooking, and seasonal water heating and room heating. Lighting needs are mostly met through kerosene, and to a certain extent by electricity, providing 2.4 per cent needs through kerosene and 0.42 per cent needs by electricity. In terms of energy consumption per capita/per day, the following figures are observed:

Cooking	::	499 kcal
Water heating	::	137 kcal
Room heating	::	256 kcal
Lighting	::	42 kcal

Energy consumption in the agricultural sector as direct inputs for cropping, sowing, irrigation and harvesting is only in terms of the manual and animal power. Agriculture is dependent on rain water or small river irrigation and traditional farming practices are followed.

Energy consumption for cottage and artisanal activities were almost non-existent in terms of commercial sources of energy. The only wheat flour mill is run on the traditional water wheel principle.

Tables 12.1, 12.2 and 12.3 give the energy consumption pattern.

TABLE 12.1: Consumption of Energy for Total Households/Year

End-use	Source	Quantity	Calorific value m kcal
Cooking	Fuelwood (Tonne)	16,200	76,140
Water Heating	Fuelwood (Tonne)	2,158	10,144
Room Heating	Fuelwood (Tonne)	2,572	12,088
Total		20,930	98,292
Lighting	Kerosene Lts.	2,93,280	2506
	Electricity (Kwhr)	4,95,232	426

TABLE 12.2: Contribution of Different Sources in Domestic Energy Consumption

Source	Percentage
Fuelwood	97.10
Kerosene	2.48
Electricity	0.42

TABLE 12.3: Energy Consumption Per Capita Per Day

	End-use	Sources	Physical Quantity	Calorific Value (kcal)	
				Total of Input	Effective Utilisation (10 per cent of total Heat)
1.	Cooking	Fuelwood (Kg)	1.05	4987	499
2.	Water * Heating	Fuelwood (Kg)	0.29	1377	137
3.	Room * Heating	Fuelwood (Kg)			
4.	Lighting	Fuelwood (Kg)	0.33	1567	156
		i) Kerosene Ltr.	0.18	153	15
		ii) Electricity (Kwhr)	0.032	27.55	27

* Being a seasonal activity, calculations based on six-month consumption.

Interaction with the Local People

The survey exhaustively covered different categories of persons to ensure their reactions for their participation in the integrated energy programme for overall development. With a view to give a clear idea, details of individual groups are mentioned below.

(a) *Gram Pradhan*

Gram Pradhans of different village panchayats indicated their interest in implementing energy related development programmes. However, they wanted total responsibility to be given to them with total accountability, but no interference from government agencies. However, they wanted the government agencies to demonstrate certain successful cases to establish the credibility and assurance about the programme.

The villagers wanted that the following should be provided in terms of facilities: (i) Financial assistance for maintaining core staff; (ii) Training; and (iii) Technical back-up.

They suggested that government gives all these facilities to its agencies then why not them?

Pradhans also suggested that the rate for electricity supplied from sources other than the conventional, should be lower as they are still in the promotional and demonstration stage. The indication was obviously for the electricity supply from the micro-hydel scheme.

(b) *Potential Entrepreneurs*

A number of young persons were identified who are willing to start income generation activities if electricity was made available to them.

(c) *Traders*

Traders dealing with the horticultural crops were extremely interested in having better facilities for crop preservation as almost 30 per cent of the crops go waste. They suggested cold storage, and proper packaging.

They also suggested that crop processing plants be set up for value addition at source.

Economic Dimension of the Study

The survey established beyond doubt the potential of integration of different sources of energy for production of electricity from decentralised sources and its availability for higher agricultural production, industrial production and corresponding improvement in income generation and employment generation. Interest of the masses from different communities and sectors indicates that people's participation can be of immense help.

It was observed that by the use of 74 kw of off-peak power from the micro-hydel scheme, a permanent employment generation of 120 persons could be established resulting in a turnover of about Rs. 2.4 million per year. Such an effort would require a capital investment of Rs. 1.675 million and a working capital of Rs. 2,48,000 only.

It will be appropriate to mention that the officials of various government agencies like the District Industries Centre, Block Development Office, Khadi & Village Industries Board etc also showed extreme enthusiasm in supporting such a programme if people themselves could implement and run them efficiently. Thus, a willingness at the local level for supporting the people's participation was observed.

Environmental Dimension

The Survey also had an environmental dimension. The respondents were quite keen on use of high efficiency woodstoves to reduce deforestation due to fuelwood consumption. There was a general awareness in the masses that deforestation has led to soil erosion and lower agricultural activities from scarce cultivable land.

An urgent need was felt for reforestation and reducing consumption of fuelwood. May be as an effect of the Chipko and other environmental movements, people are quite alert to use of their energy resources in an environmentally sound fashion.

Energy - Development Interaction

The survey indicated that the masses were interested in utilisation of fuelwood and kerosene for the domestic needs and electricity mainly for income generating needs and then for domestic needs.

Interaction with Local Voluntary Organisations

Eight different voluntary agencies were identified involved in rural development programme. Discussions were held with a number of individuals and their reactions, collectively can be described as follows:

(a) The most preferred energy options should be related to improvement in energy use efficiency and local resource based electricity generation for income generating activities. Examples were referred to improvement in efficiency in wood burning and power generation from micro hydel scheme.

(b) There was a close agreement on suitability of a number of technologies for use in hilly areas. Domestic solar water heating, passive architecture, solar refrigeration, photovoltaic power for telecommunication, and high efficiency woodstove were seen as promising technologies that can be immediately transferred.

(c) Environmental issues were considered important when energy resource identification was carried out. For example, use of gasifiers based on fuelwood was not supported as that would lead to deforestation.

(d) The voluntary agency perception was that providing energy at a reasonable cost should be the primary issue for ensuring adequate supply and consumption within the local purchasing capacities.

(e) Following factors were considered significant elements consumer to decide on energy options.

— Short-term cost of energy;
— Availability;
— First cost of equipment;
— Running cost; and

— Reliability.

(f) Voluntary Agency participation to be effective in implementation of various energy programmes oriented to development should involve government agencies providing core support, as the voluntary agencies undertake the work which otherwise would have been executed by the government agencies.

(g) Voluntary Agencies work as the best interface between the government agencies and the common man.

Strategy for Voluntary Agency Involvement in IREP Implementation

On the basis of the observations made during the Case Study, the following strategy can be considered for participatory action of the masses in implementation of I.R.E.P.

Type of Activities

People's participation can be effective in the following issues of IREP.

(a) Creation of an awareness of various technologies amongst the masses for their ready acceptance in case of proven and reliable energy technologies. Public acceptance of new technologies will never be uniform. Such a situation may distort the perception of government agencies. Effective participation of voluntary agencies will ensure two way communication;
(b) Survey of 'resources' and 'needs' to match the supply of technologies and devices;
(c) Project preparation and finalisation with the government funding agencies;
(d) Implementation of schemes; and
(e) Monitoring and evaluation of projects.

Facilities Required

Voluntary agencies and other people's groups will require the following facilities:

(a) Support for core staff;
(b) Training of personnel;

(c) Technical back-up;
(d) Institutional finance for related activities;
(e) Technology packages for income related programmes based on introduction of I.R.E.P.

Advantages

People's Participation will have the following advantages in implementation of I.R.E.P.

(a) Masses will consider the implementation of programmes as their own in view of the participatory nature of projects. Results will be most cost-effective.
(b) Voluntary agencies are observed to implement programmes in a low overall financial expenditure and therefore will have more projects for the same funding.
(c) Nodal Agencies after off-loading their implementation responsibilities will have more time for planning, supervision and evaluation of programme.
(d) The data collection will be much more true to the ground realities when collected by the local organisations than when it is collected by outside government agencies.
(e) People's organisation will be the most important motivating and driving force for the IREP.

Conclusion

The Cost Study indicates that 'People's Participation' in I.R.E.P. can be the most cost-effective route for integration of energy with development in the rural areas.

REFERENCE

Study on Socio-Economic Feasibility of Renewable Energy based systems in the hilly areas of Uttar Pradesh - for income generation, Indian Association for the Advancement of Science (1991).

13

TECHNO-ECONOMICS OF SOLAR GREENHOUSE FOR THE COMPOSITE CLIMATE

Introduction

Solar greenhouse has been a major application in Indian agriculture now poised for a major breakthrough, both for the local market as well as for export. Climate variation is the major bottleneck for biological plant growth, and therefore, quality supply to export market becomes constrained.

Greenhouse technology has been identified as an important technology that helps in higher growth and higher productivity levels in agriculture. How can these levels be achieved? Greenhouse can effectively be utilised to enhance growth and productivity and maximise returns from the field. Greenhouse application has originated in the cooler climate of northern countries; and accordingly been disseminated for use of greenhouse applications for collection of heat for vegetable and fruit production as well as warming the houses.

Composite Climate of Delhi

The peculiar situation of Delhi is that the temperatures fluctuate from 1°C in the winter nights to 48°C on a hot summer day. Average night time temperatures in winter is only 8°C. Therefore, the real challenge for developing a solar greenhouse with the composite climate lay in bringing down the summer time temperature from the ambient of 45°C

Paper submitted to the International Solar Energy Society Convention 1995 at Harare, Zimbabwe, during 9-16 September, 1995.

to around 27°C; and for winter application efforts would be to raise the temperature from 8 to 10°C ambient to 27°C.

Design Considerations

It was observed that such an approach can help in production of the final product or intermediary like seedlings and sprouted seeds for quality production of final products compared to normal conditions. Greenhouse was also identified to be suitable for all the three crops - paddy seed sprouting; seedling production for vegetables and fruit production. Specific applications were identified and then the greenhouse was designed.

Initial application of the solar greenhouse was the starting point for considering the design parameters. A multi-disciplinary group consisting of physicists, biologists and agricultural engineers was set up to design. It was decided that instead of using the greenhouse for total production it would be better if we use the greenhouse for developing an intermediary like the seedlings. Thus, the summer hot months of May, June and July were initially suggested for production of seedlings so that they were ready for transplantation at the onset of the monsoon, thereby resulting in vegetable production when high price was fetched in the market.

Design of the Solar Greenhouse

Solar greenhouse was designed as a demonstration greenhouse to study its performance under the composite climate of Delhi in the year 1993. Wind direction in Delhi is west to east and, therefore, the greenhouse was designed with its longer surface in the east-west direction. The south facing wall was made with toughened glass so that solar radiation could be absorbed and the north wall was made out of the particle board to avoid solar radiation during the summer months. Solar greenhouse was designed with a floor size of 26 ft x 22 ft and a vertical height of 12 ft. Cross- flow of wind was organised in the longitudinal direction with wind blowing from west to east. Evaporative cooling arrangement was made with cooling pads on the west side covering an area of 42 sq ft and with 3 exhaust fans on the east side-2 of them of 24 inch diameter, and 1 of 18 inch diameter so that the air could be exhausted over the plants at a surface speed of less than 1 metre/second. (Seedlings grown inside wilt at a surface speed higher than 1 metre/second). Also air changes at 2 per minute or 120 per hour had to be attained to carry away the heat and contain the temperature to required limits (in conjunction with evaporative cooling arrangement) and to have the best photosynthesis effect. 3 fans had to be chosen so that the air is uniformly exhausted from all the directions. Cooling pads provide required humidity inside the greenhouse. For monsoon period, when the humidity is very high, water supply to cooling pads was stopped and only air changes at 2 per minute were attained. During the winter months, cooling pads were completely sealed so that no air could come through that end; and the air change rate was reduced to 45 per hour.

Uses of Solar Greenhouse

Following uses of the solar greenhouse were carried out during the experiment:

Vegetable Seedling Nursery

Seeds were sown for the first time in the solar greenhouse on May 3, 1993 and fully grown saplings were taken in 5 weeks and were transplanted in the field. Quality of seedlings was observed to be very good and survival in the nursery was almost 90 per cent. In comparison, the survival in the nursery grown in the open field under the tree was less than 25 per cent, and that under the shade was about 40 per cent. Such plants were then transplanted into the field with a 10 per cent survival rate and the productivity levels were much higher. In case of hybrid tomato, it was observed that the productivity level of 40 tonnes per hectare was attained in the field; and a number of plants carrying upto 45 tomatoes per plant were observed. Other vegetable seedlings that were grown were cucurbits, brinjal, cauliflower, cabbage, ginger and chilly.

Paddy Sprouting

Paddy nursery was made in the open field for transplantation in end of June 1993. As per the practice, seed is first moistened with water and kept under humid conditions for two days to effect germination before planting in the nursery. As the temperatures were very high in the months of April and May 1993, germination of seeds in the open was very poor. Subsequently, experiments were carried out to germinate the seed in the solar greenhouse. 100 per cent germination with extraordinary sprouting was observed which helped in raising quality nursery for paddy crops for various varieties of high value added rice.

Observations on Performance

Specific observations made on the solar greenhouse are as follows:

(i) Growth of plants inside the greenhouse is better during its germination and initial period of growth.

(ii) Growth of seedlings inside the greenhouse was much faster than the growth of the seedlings in open nursery.

(iii) Growth of seedlings in polysacks provided better control with regard to watering, manuring and transplanting to the open field.

(iv) Plants under covered north wall grow taller and leaner as compared to those under the south side with glass glazing. This is mainly due to uniform light being available on the south-end.

(v) Growth of all the plants on transplantation to the open field is same and is much better than from the nursery grown in the open field. Thus, even seedlings growth under north wall that appeared lean and tall have undergone production in the same fashion as the other healthy appearing plants.

(vi) During the summer months of June when electricity breakdown took place for about one week at a stretch, the plant leaves burnt on its periphery. Although, the temperature inside was kept equal to the ambient temperature by keeping all doors and windows open but the plants did not receive the surface air movement as in the open field. This is an indicator to an inherent cause of concern while using greenhouse in hot climate with erratic power supply. However, subsequently on resumption of power supply and maintenance of air flows, the plants could regain their green.

(vii) Plant growth inside the greenhouse was an optimum for temperature around 27°C. A fluctuation (of 2°C) in temperature on either side could adversely affect plant growth.

(viii) Seedling transplantation in the month of June/July and October/November (end) have been without any problem and 100 per cent transplantation survival has been reported.

Precautions Evolved for the Greenhouse Application

Certain precautions were identified on the basis of observations made in the field.

(i) Uninterrupted electric supply to the greenhouse is extremely necessary.
(ii) Light intensity inside the greenhouse should be uniform for uniform plant growth.
(iii) Soft water should be used for irrigation of seedlings.
(iv) Organic manure only should be used, and, if necessary, only small doses of chemical fertilizers be used in the greenhouse.
(v) Insecticides and herbicides should not be used.
(vi) Agricultural practices suited to greenhouse should be strictly followed.

Economic Returns and Cost Benefit of Solar Greenhouse

Data collected on solar greenhouse operation for one year has revealed that an overall agricultural productivity and thereby the returns through value addition to the crops increased tremendously.

In one year 8 crops of the saplings were grown inside the greenhouse after giving an intervening period of 1 week after every crop. At one time about 5,000 plants were grown inside the greenhouse and transplanted onto the field. These 5,000 plants covered an area of 1 hectare. If solar greenhouse was not used and 1 hectare of tomato was grown then the production would have been only a maximum of 5 tonnes per hectare for one crop. With the use of hybrid seeds and saplings grown in a solar greenhouse, productivity of 40 tonnes per hectare of tomato was observed. The tomato growth was early and as such there was a price advantage also. Maximum price for the seasonal tomato in 1993 in the wholesale market was about Rs.10 a kg, whereas off-season hybrid tomato sale price, in the wholesale market was about Rs. 20 a kg. Thus, a possibility of improvement in income upto 10 times was observed. This shows also a remarkable difference in the production because of use of solar greenhouse, better growth of seedlings and early transplantation into the field.

Similarly, 8 different crops grown in a year showed a benefit of almost Rs. 50,000 per crop, aggregating to almost Rs. 400,000 in a year. The cost of the solar greenhouse was about Rs. 100,000 and the O&M costs were about Rs. 20,000. Inspite of all these additional investments, the benefits completely outweigh the expenditure and as a matter of fact the cost of the solar greenhouse was recoverable in less than a year.

Conclusion

The experiment on solar greenhouse in the composite climate of Delhi was a small step towards field demonstration of the technology. It also opened new avenues for solar greenhouse intervention for value addition in agriculture. The experiment also demonstrated that solar greenhouse technology is highly profitable when used as an intermediary.

14

INDIA'S FIRST SOLAR CHICKEN BROODER

Introduction

A 1200 bird solar chicken brooder was indigenously designed and operated by the Indian scientists for the first time in 1993. This multidisciplinary project was taken up on the International Sun Day, May 3, 1993 and completed on May 3, 1994. Data was collected for the first nine months of operation till December 31,1994. Its successful operation has justified the multidisciplinary approach.

Temperature control in a chicken brooder is essential as infant chicks are very sensitive to the extreme temperatures of summer (hot) and winter (cold) experienced in Delhi. In the absence of measures to control temperature inside the building, the mortality rate of chicks can be high, accompanied by non uniform growth. Operating cost of active thermal control system, is very high and the system also requires trained manpower for operation. On the other hand, passive systems can be integrated with the building and do not require fossil fuel or trained manpower to operate. The design adopted for the chicken brooder was hybrid - having passive as well as active energy components. Active components are incorporated so as to make the system more effective during specific periods of time when passive components alone are not sufficient to meet the thermal control requirements.

The solar chicken brooder incorporated modern poultry concepts of breeding under controlled temperatures. In view of the mixed climate of Delhi, provision has been made for both the heating and the cooling to take care of the 24 hour cycle. Comfort conditions have been identified

---

Paper presented at the International Symposium on Solar Energy, organised by IEEE at Miami, USA in June, 1995.

and maintained as is done in the air-conditioning for human beings.

Design Concept

The geographic location of Delhi and the meteorological data available for the number of years indicates the composite climate requiring extra care separately for winter, summer and monsoon seasons. Of all the three seasons the most difficult is the monsoon season as the level of relative humidity is high and requires much higher air changes to provide comfort conditions. Whereas for the winter months the air changes are only at the rate of 3 per hour the corresponding rate of air changes for the monsoon months is 10-12.

These comfort conditions are created by using the hybrid solar passive/active system in combination with the air washer. Heating and cooling loads have been effectively reduced by using low cost rat trap bond technology for wall construction and hollow block roof with mud fusca topping.

Basic Parameters

Basic parameters for the concept of design were laid down as follows:

(i) for temperature conditions by the poultry expert;
(ii) for solar insolation and wind speed conditions by solar experts, and;
(iii) for availability of building materials by the architect.

These parameters differ with the age of chicks and are as follows:

(i) Temperature	(a) 95° F (0-2 weeks)
	(b) 90° F (3-4 weeks)
	(c) Room Temp. (5-8 weeks)
(ii) Relative Humidity	20-60 per cent
(iii) Living	0.5 sq ft/bird (0-4 weeks)
	1.0 sq ft/bird (5-10 weeks)

Consideration of Internal and External Factors

For the incorporation of passive components in the building, both internal and external factors have been taken into consideration.

External Factors

External factors are mainly climate and weather characteristics. These include:

- — Solar Radiation Inputs (both direct and diffuse);
- — Wind Conditions;
- — Relative Humidity; and
- — Precipitation

All these factors are a function of latitude, altitude, topography and land/water relationship and have been given due consideration.

Internal Factors

Internal factors crucial for designing include the following:

- — Internal thermal load of the occupants.
- — Floor area.
- — Artificial and day light conditions
- — Number of openings.

Built in Characteristics for Comfort Conditions

For providing comfort conditions for the chicks, provision has been made for the following:

(i) Summer control on solar radiation by shading; Evaporative cooling and ventilation during summer.

(ii) Monsoon ventilation in high humidity conditions.

(iii) Direct solar heating with trombe wall and indirect heating from solar heat and water storage during winter.

Maximum Heat Gain in Winter

Since the winter time temperature in Delhi reach a low of 1°C during night and the average day time temperature is about 8-10°C, it was considered necessary to provide heating arrangement during winter through passive and active systems. Useful day time solar gain depends on the glazed area exposed to the sun, type of glazing and orientation of the building. For Delhi the South-East and the South-West directions are for heat gain during the winter months. Utilising the sun chart and the climate data, (and considering the existence of the building in the vicinity of the solar chicken brooder), the orientation has been so decided that the south wall (wall facing the south) is at an angle of 25 degrees tilt in azimuth towards west. The ideal situation would have been normal to the south. However, it has been observed that the heat collection loss is only 7 per cent or negligible for that orientation.

Minimum Heat Gain in Summer

This has been built into the design so that the cooling load is minimum. To attain such a condition properly designed solar protection over closed windows in the form of shading and overhangs has been provided. In view of the rotation of the south wall and shade protection in different months, the initial design shade of 1 metre overhang was reduced to only 10 cms at the roof level. To provide protection from rain the window sill was so designed that the windows were actually about 27 cms inside the wall.

With a view to keeping the temperatures low in summer, (except for the south wall), a 2 metre open space with hollow block covering at the roof level was provided on three sides of the solar chicken brooder. This prevented direct sunlight falling on the three walls.

The central passage of the building is in the east-west direction with sufficient openings through the passage right for cool draft to flow through the passage during the summer months and keeping temperature low. During winter months this passage was closed at one end, except as desired for air changes.

Design Details

The basic building design adopted was of high thermal mass. In the present case, double wall construction has been used; side walls have perfect insulation because of brick walls on either side and air gap in between walls; and south wall has a solid mass. Roof has hollow blocks with adequate insulation due to double walled construction and the air in between the two layers. Further, the mud fusca topping and the brick tiles gives optimum insulation. In addition, some solar radiation on roof is cut off by solar collectors, piping, storage tank and the air washer equipment installed on the roof.

South wall forms the solid thermal mass and is convective. It has been used to gain both convective as well as radiative heat transfer during the heating period. A gap of 7 cm between the wall and the glazing has been provided to ensure good flow of air across the face of the trombe wall. Gap at the top of the glazing is only 5 cm which allows for quicker air circulation during sunny period. 4 mm thick float glass has been provided for the glazing. Three openings of 40 cm x 10 cm have been provided on the top and the bottom of the south wall as the air vents for air circulation. In order to prevent reverse cooling during winter night, the wall has been extended below the ground level, which provides a trap for the cool air. Air vents at the bottom have been elevated above the ground so as to prevent chicks from entering the air space between the glazing and the wall. In addition, the openings have been covered with wire mesh. Movable louvres have been provided at the top and the bottom of the vents so as to prevent air circulation when not needed.

Direct impingement of solar radiation on the southern wall has been so designed as to allow solar heating of wall upto the end of April.

Active solar water heating system of 600 litres per day capacity, at 80°C, has been installed to supplement the heating by passive arrangement. Hot water to air heat exchanger has been built into the storage tank and hot air in various compartments is designed to recirculate during winter months so that the heating load is low. Induction of fresh air continuously would have increased heating requirement considerably.

An air washer, i.e., an evaporative air cooling arrangement, has been provided to keep the temperature at desired levels during hot summer months when the ambient goes upto 40°C at Delhi. The blower used with air washer is also used for circulation of hot air when heating is required. This provides low cost heating and cooling arrangement.

Design Calculations

Following design calculations were made for solar heat absorption by the trombe wall and the solar water heating system in the month of December (as this was considered to be the coldest period of the year). This has been calculated on monthly average daily value.

Assumptions

It is assumed that the energy absorbed by the trombe wall, net of heat losses from it through the front glazing, is eventually used in heating the enclosed space:

(a) either by conduction to the inner surface and then convection and radiation to the enveloping air; or
(b) carried by air through vents in the trombe wall.

The assumptions are, therefore, summarised as below:

(i) Thus, the effective contribution of trombe wall is calculated as the difference between the energy absorption by it and the heat losses from it to the surroundings.
(ii) Ambient temperature is assumed to be 15°C during sunshine hours (8 hours) and 10°C during off-sunshine hours (16 hours in a day) for the month of December.
(iii) Ground temperature is assumed to be 18.5°C
(iv) Solar heat gain contribution from the roof and other sides have been assumed to be negligibly small.

Calculation for heat absorption by Trombe wall and heat collection by solar water heating system.

—	Extra terrestrial solar radiation on a horizontal surface for a whole day in the month of December (8 hours of sunshine).	: 5.767 kwh/sqm
—	Daily sum of direct solar radiation on a horizontal surface.	: 2.754 kwh/sqm
—	Daily sum of diffuse solar radiation	: 1.196 kwh/sqm
—	Daily sum of global radiation	: 3.950 kwh/sqm
—	Daily sum of global solar radiation on a clear day under average conditions.	: 4.231 kwh/sqm
—	Clearness index	: 3.950% 0.685 5.767

Conclusion

Birds were reared (as per their genetic characteristics) at different temperatures for a period of 8-10 weeks to grow them to a uniform weight of 2.0 kgm. Growing them under controlled temperature for the first 4 weeks and then at room temperature was another new concept to grow hardy stock.

Following 2 characteristics need be appreciated

(i) Simple operational procedures made the system easily operational by the least technically trained manpower.
(ii) Production of sequential crops of chicken has established the financial viability of the system.

Solar Chicken Brooder is the First Pilot Plant in India and has undergone vigorous trials. Technical viability is proven and initial results indicate financial viability also. Its technical and financial viability under different climatic conditions has still to be proven. This small beginning can lead to revolutionary growth of chicken brooding industry.

(NOTE: The Paper was authored jointly with Dr. S.K. Sharma; Prof. T.S. Narayanaswamy; Mr. Atam Kumar and Mr. Udayan Choudhary)

ANNEXURES

ANNEXURE-I

GLOBAL ENERGY PERSPECTIVES 2050 AND BEYOND

Joint IIASA/WEC Study Report 1998

AUTHORS : IIASA
Nebojsa Nakicenovic, IIASA
Arnulf GrÜbler, IIASA
Alan McDonald, IIASA
and
Michael Jefferson, WEC

Introduction

The support paper summarizes the results of a five-year study conducted jointly by the International Institute for Applied Systems Analysis (IIASA) and the World Energy Council (WEC). The starting point is the three alternative cases and one variant of future economic and energy development presented in 1993 in Energy for Tomorrow's World. There, the principal analysis covered the period through to 2020 with some extensions to 2100. This study describes three cases of alternative energy futures that diverse into a total of six scenarios, and their implications for 11 world regions. The objective was to examine more thoroughly the period beyond 2020, where the real potential for change lies. To that end, the study integrates near-term strategies through 2020 with long-term opportunities to 2050 and beyond; analyzes alternative futures with a unified methodological framework using formal models and databases to ensure consistency and reproducibility; incorporates a dynamic treatment of technological change; harmonizes regional

(The Authors presented the following Support Paper as Summary of their Report at the 17th Congress of the WEC at Houston, during September 13-18, 1998.)

aspirations with global possibilities; and takes account of new developments and new data since Energy for Tomorrow's World appeared in 1993.

The study was conducted in two phases; The first, from 1993 to 1995, developed the six scenarios for 11 world regions and analyzed their implications. The results were presented in the 1995 IIASA-WEC report Global Energy Perspectives to 2050 and Beyond. The second phase, from 1995 to 1998, was decoded to an extensive review of the study assumptions, results, and implications for the 11 world regions. Teams of regional experts and reviews, totalling over 100 individuals, were convened to provide more thorough regional assessments and alternative perspectives. The final results, incorporating the regional reviews plus an updated, expanded, and more detailed presentation of the six scenarios than in 1995, are presented in a new book entitled Global Energy Perspectives, published this September by Cambridge University Press.

Overview of the Scenarios

The study's six scenarios are divided into three cases. The principal focus for all cases is on the period up to 2050, but results are also presented out to 2100. In brief, Case-A presents a future of impressive technological improvements and consequent high economic growth. Case-B describes a future with less ambitious, though perhaps more realistic, technological improvements, and consequently more intermediate economic growth. Case-C presents an ecologically driven future. It includes both substantial technological progress and unprecedented international cooperation centered explicitly on environmental protection and international equity. Key characteristics of the three cases are given in Table 1.

All three cases provide for substantial social and economic development, particularly in the developing world. They provide for improved energy efficiencies and environmental compatibility, and thus for associated growth in both the quantity and the quality of energy services. Across all three cases, the structure of final energy develops in the same way and energy intensities improve steadily. To facilitate comparisons among the three cases, all share the same central demographic baseline assumption in which global population grows to 10 billion (109) by 2050 and to nearly 12 billion by 2100. All have been checked for internal consistency with the aid of formal models and

TABLE 1: Summary of the three cases in 2050 and 2100 compared with 1990

	Case		
	High growth	Middle course	Ecologically driven
Population, billion			
1990	5.3	5.3	5.3
2050	10.1	10.1	10.1
2100	11.7	11.7	11.7
GWP, trillion US (1990) $			
1990	20	20	20
2050	100	75	75
2100	300	200	220
Global Primary energy intensity improvement, per cent per year	Medium	Low	High
1990 to 2050	-0.9	-0.8	-1.4
1990 to 2100	-1.0	-0.8	-1.4
Primary energy demand, Gtoe			
1990	9	9	9
2050	25	25	25
2100	45	35	21
Resource availability			
Fossil	High	Medium	Low
Non-fossil	High	Medium	High
Technology costs			
Fossil	Low	Medium	High
Non-Fossil	Low	Medium	Low
Technology dynamics			
Fossil	High	Medium	Medium
Non-Fossil	High	Medium	High
Environmental taxes	No	No	Yes
CO_2 emission constraint	No	No	Yes
Net carbon emissions, GtC			
1990	6	6	6
2050	1-15	10	5
2100	6-20	11	2
Number of scenarios	3	1	2

Abbreviations: GWP = gross world product; Gtoe = gigatonnes oil equivalent; CO_2 = Carbon dioxide; GtC = gigatonnes of carbon

Source: WEC/IIASA Study on Global Energy Perspectives, 1996.

through a lengthy regional expert review process.

Case A includes three high-growth scenarios that address key developments in energy supply. They vary principally in the future they envision for coal, on one side, and nuclear and renewables, on the other. In Scenario A1, there is high future availability of oil and gas resources. Dominance of oil and gas is perpetuated to the end of the 21st century. At the other end of the spectrum, Scenario A2 assumes oil and gas resources to be scarce, i.e., limited to known reserves, resulting in a massive return to coal. Finally, in Scenario A3 rapid technological change in nuclear and renewable energy technologies results in a phase out of fossil fuels for economic reasons rather than due to resource scarcity.

Case B has a single "middle-course" scenario. It incorporates more modest estimates of economic growth and technological development, and the demise of trade barriers and expansion of new arrangements facilitating international exchange. Compared with the Case A scenarios (and the Case C scenarios), it is more "pragmatic", which is its main appeal. Case B manages to fulfil the development aspirations of the South, but less uniformly and at a slower pace than in the other cases. For regions such as Africa, progress is painfully slow.

Case C is the most challenging. It is optimistic about technology and geopolitics, but unlike Case A, it assumes unprecedented progressive international cooperation focused explicitly on environmental protection and international equity. It includes substantial resource transfers from industrialized to developing countries, spurring growth in the South. These resource transfers, which recycle funds from the OECD to developing countries, reflect stringent international environmental taxes and incentives to reduce carbon emissions in 2100 to 2 gigatons of carbon (GtC) per year, one-third of today's level. In Case C, nuclear energy is at a crossroad. Two scenarios are included. In Scenario C1, nuclear power proves a transient technology that is eventually phased out entirely by the end of the 21st century. In Scenario C2, a new generation of nuclear reactors is developed that is inherently safe and small scale ù 100 to 300 megawatts electric (MWe) installed capacity ù and finds widespread social acceptability.

Across the three cases, the primary energy needs increase up to threefold by 2050 and up to fivefold by 2100. The six scenarios diverge in terms of the contributions of individual primary energy sources ù what percentage of primary energy is supplied by coal, what percentage by oil, and so forth.

Because of the long life times of power plants, refineries, and other energy investments, there is not enough capital stock turnover in the scenarios prior to 2020 to allow them to diverge significantly. But the seeds of the post-2020 divergence in the structure of energy systems will have been widely sown based on by then RD&D efforts, early market deployment, intervening investments, and technology diffusion strategies. It is these decisions between now and 2020 that will determine which of the diverging post-2020 development paths will materialize.

After 2020 all scenarios move away from their current reliance on conventional oil and gas. This transition progresses relatively slowly in Scenario A1 where oil and gas are plentiful. In Scenario A3 and Case C, it progresses more rapidly due to faster technological progress (Scenario A3) or because of energy and environmental policies favouring non-fossil fuels (Case C). In Scenario A2 and Case B, the transition away from oil and gas includes an important contribution from coal, whose long-term market share after 2050 ranges between 20 and 40 per cent. Nonetheless, little of this coal is used directly. It is instead converted to the high quality energy carriers (electricity, liquids, and gases) demanded by the high-income consumers of the second half of the 21st century.

Despite the divergence in primary energy structure, the pattern of final energy use is remarkably consistent across scenarios, showing a continuing trend toward energy reaching the consumer in ever more flexible, more convenient, and cleaner forms.

All three cases reflect a continuing pervasive shift from energy used in its original form, such as traditional direct uses of coal and biomass, to elaborate systems of energy conversion and delivery. This shift continues in all cases, leading to ever more sophisticated energy systems and high quality energy carriers. A second profound transformation is the increasing degree to which energy is delivered by dedicated transport systems, such as pipelines and networks. This development enhances trade possibilities and promotes similar end-use patterns across regions with fundamentally different primary energy supply structures. Finally, changes in final energy patterns reflect the changes in economic structure presented in the scenarios. As incomes increase, the share of transport and residential and commercial applications also increases.

Regional Reviews

The second phase of the study focused on regional implications of the global scenarios. It had two parts. First, more than 100 regional experts organized into 11 groups corresponding to the study's 11 world regions developed detailed regional assessments of the scenarios. The key result was that the detailed regional reviews confirmed the overall conclusions from the initial global analysis. Detailed findings of the regional reviews are therefore presented as part of the study's overall conclusions in the next section.

Second, regional experts were polled to identify the relative importance of various energy-related concerns within regions and across regions. The groups were asked to score eight issues central to energy policy debates : population growth, demand for commercial energy services, technology, financing, institutional deficiencies, efficiency and conservation, local environmental concerns, and possible climate change. The issue of local environment concerns was added to the six issues listed in the 1992 WEC poll, and one of the 1992 categories, energy need and population growth, was split into two. Issues were scored according to the following scale : 1 = very important; 2 = important; 3 = of concern; and 4 = of no concern. The results are shown in Table 2.

In many respects, Table 2 confirms conventional wisdom. But, it also holds some surprises. Considering all 11 regions together, institutional deficiencies and financing are the top concerns. Both issues have moved ahead of efficiency and conservation, the top ranked issue in the 1992 poll. Even North America (NAM), often considered an institutional role model, ranked institutional deficiencies as a number one priority. For all 11 regions as a group, possible climate change is at the bottom of the list, as it was in 1992. In general, it is of more concern to the OECD regions than to either the developing or reforming regions. In all regions, local environmental concerns are ranked higher than or equal to possible climate change. Western Europe (WEU) lives up to its green image, being the only region to give local environmental concerns a number one ranking. NAM is unique in scoring technology as a number one issue.

While population growth is generally of greater concern in developing regions than in reforming or OECD regions, it was also ranked relatively high by PAO. However, the reason is Japan's concern about too little population growth, exactly the opposite of the developing regions' concerns about too much population growth. PAO's top ranking for efficiency and conservation, and NAM's top ranking for technology reflect the two regions' positions on the cutting edge of these fields. Given the importance of technology, efficiency, and conservation in the scenarios, the continuing high priority given to these issues is reassuring.

Conclusions

World Energy needs will increase

World population is expected to double by the middle of the 21st century, and economic development needs to continue, particularly in the South. According to the scenarios of this study, this results in a 3 to 5-fold increase by 2100. By 2100, per capita income in most of the currently developing countries will have reached and surpassed the levels of today's developed countries. Disparities are likely to persist, and despite rapid economic development, adequate energy services may not be available to everyone, even in 100 years. Nonetheless, the distinction between "developed" and "developing" countries, in today's sense, will no longer be appropriate. Primary and final energy use will grow much less than the demand for energy services due to improvements in energy intensities. We expect a 1.5 to 3-fold increase in primary energy requirements by 2050, and a 2 to 5-fold increase by 2100.

For the perspective of the developing regions where energy demand will grow the most, there was a clear "winner", the high growth Case A. The ecologically driven Case C did not find widespread appeal, despite its explicit focus on international equity including massive resource transfers from North to South. Thus, in Asia and several other developing regions, such as Latin America and the Caribbean (LAM) and the Middle East and North Africa (MEA), there was a clear preference for the vigorous development represented by Case A, characterized by high rates of economic growth with developing countries quickly "catching up". There were differences of opinion about how quickly the high growth path of Case A could unfold, with regional aspirations generally favouring short to medium term economic growth rates even above those of Case A.

The recent experience of Central and Eastern Europe (EEU) and the newly independent state of the former Soviet Union (FSU) demonstrates that economic growth can by no means be taken for granted. The recessions in EEU and FSU during the 1990s have been deep and severe. The overall paths of the recessions were anticipated quite closely in all scenarios, but, unfortunately, both the scenarios and recent developments suggest that the process of recovery will be a long one. In EEU, per capita GDP recovers to 1990 levels by 2010. In FSU, 1990 levels of per capita income are reached again between 2010 and 202C. However, by that time the same amount of GDP per capita as in 1990 represents a considerable increase in personal income and welfare as a

TABLE 2: Regional poll results. Areas of regional concerns are ranked as follows :
1 = very important, 2 = important, 3 = of concern, 4 = of no concern

Region	Population growth	Demand for commercial energy supplies	Technology	Finan-cing	Institu-tional defici-encies	Efficiency and conser-vation	Local environ-mental concerns	Possible climate change
NAM	4	3	1	4	1	3	2	2
LAM	2	2	2.5	1.5	2	2	2	2.5
WEU	3.5	3	2	3	2	2	1	2
EEU	4	2.5	2.5	1.5	2	1.5	2	3.5
FSU	4	4	3	1	1	2	3	4
MEA	1	3	2	1	1	3	3	4
AFR	1.5	3	2.5	1.5	1.5	3	2.5	3
CPA	1	2	2	1	1	1	2	2
SAS	2	1.5	2	1.5	1	2	1.5	4
PAS	3	1	2.5	1.5	2.5	2	2.5	3.5
PAO	2	2.5	1.5	2	2.5	1	1.5	1.5

result of the overall economic shift away from smoke-stack industries toward consumer goods and services.

Energy intensities will improve significantly

As individual technologies improve, conversion processes and end-use devices progress along their learning curves, and as inefficient technologies are retired in favour of more efficient ones, the amount of primary energy needed per unit of economic output ù the energy intensity ù decreases. Other things being equal, the faster the economic growth, the shorter the obsolescence time, the higher the turnover of capital, and the greater the energy intensity improvements. In the six scenarios of this study, improvements in individual technologies are varied across a range derived from historical trends. Combined with the economic growth patterns of the different scenarios, the average global reductions in energy intensity range between 0.8 per cent per year and 1.4 per cent per year. These bracket the historical rate of approximately 1 per cent per year and cumulatively lead to substantial energy intensity decreases. Some regions improve faster, especially where current intensities are high and economic growth and capital turnover are rapid.

The regional reviews endorsed the overall conclusions on energy intensity improvements, although there was a large difference of opinion across the regions on the rate of future improvements. The general pattern was that regions with large domestic resources and less auspicious economic growth over the past decade (MEA, LAM and NAM) tended to be more conservative regarding future energy intensity improvements. In contrast, regions with more limited domestic resources that have recently experienced rapid economic growth and energy intensity improvements (CPA and SAS) considered further rapid improvements to be not only possible but necessary to sustain rapid growth.

This pattern is consistent with the basic historical relationship between economic growth and energy intensity improvements incorporated in the scenarios ù namely, other things being equal, economic growth rates depend on the distance to the productivity frontier and on capital turnover rates, and energy intensities improve more quickly with faster economic growth and more rapid turnover of capital_stock. Thus, growth is faster in regions with low per capita income and low labour stock. Thus, growth is faster in regions with low per capita income and low labour productivity than in regions with high incomes and productivity. Similarly, energy intensity improvements are faster in regions with high energy intensities than in regions closer to the energy productivity frontier.

Resource availability will not be a major global constraint

The resource scarcity perceived in the 1970s did not occur as originally assumed. With technological and economic development, estimates of the ultimately available energy resource base will continue to increase. A variety of assumptions about the timing and extent of new discoveries of fossil energy reserves and resources (conventional and unconventional), and about improvements in the economics of their recoverability, are reflected in the range of scenarios reported here. All, however, indicate that economic development over the next century will not be constrained by geological resources. Regional shortages and price increases can occur, due to the unequal distribution of fossil resources, but globally they are not a constraint. Environment concerns, financing, and technological needs appear more likely sources of future limits. The short-term volatility of international politics, speculation, and business cycles will periodically upset the long-term expansion of resources.

It is noteworthy that no regional review voiced concerns, so dominant even a decade ago, about imminent resource scarcity. Another change in emphasis ù in agreement with the global analysis ù is that resource availability is no longer seen as geologically preordained. It is viewed more as a function of the incentives and policies put in place for exploring and developing resources, constructing the necessary long-distance transport infrastructures, and, above all, attracting capital to energy investments. All are necessary to translate potentially vast geological resources into economically and technically recoverable reserves.

Import dependence and export possibilities were important concerns in several regions. The MEA regional review voiced concern about medium-term export potentials, particularly should regional energy demands grow much faster than anticipated in the scenarios, thus "crowding out" possible export volumes. Similar trends were projected in the PAS regional review. Concerns about import dependence were particularly strong in WEU, a regional traditionally heavily dependent on energy imports. The NAM regional review expressed concern about coal-intensive developments in Scenario A2, with its enormous synfuel production and exports. This concern raises the question of whether NAM would be prepared to invest in the massive coal development and synfuel exports associated with Scenario A23. More generally, with the exception of CPA and SAS, coal resources are concentrated in the affluent North, regions that may not necessarily depend on revenue generated from energy exports.

Overall, the regional reviews concurred with the global perspective that long-term revenues for energy exporters will remain solid. Oil export revenues for MEA are high in all scenarios : upto US $ 300 billion by 2020 (and even higher thereafter) in Case A, some US $ 200 billion in Case B and US $ 140 billion in Case C. The region need not fear revenue losses due to policy measures aimed at reducing CO_2 emissions in industrialized countries. Even in the ecologically driven Case C, combined oil and gas export revenues are higher than in 1990. For FSU, gas exports consistently increase in all scenarios through 2020, reaching around 300 billion cubic meters per year with revenues increasing to about US $ 50 billion, five times the 1990 value. Beyond 2020 exports and revenues increase further in Scenarios A2 and C1 and Case B, with revenues exceeding US $ 100 billion, 10 times the 1990 value.

Quality of energy services and forms will increasingly shape future energy systems

The regional reviews strongly endorsed the global conclusion that the historical trend by which final energy forms become cleaner and more grid-dependent as affluence grows can continue for a wide range of possible future primary energy structures. Solid fuels such as traditional biomass and coal in residential and commercial applications are gradually phased out in all scenarios. Grid-dependent fuels such as electricity, natural gas, and eventually also hydrogen are phased in. Although the speed of the transition varies, the trend everywhere ù in global and regional perspectives alike ù is toward cleaner, more convenient, and more flexible end-use fuels. The consistency between the regional reviews and the global analysis in their emphasis on the structure of final energy use is particularly important, as this feature of the study represents a conceptual and methodological advance over traditional scenario studies focused solely on energy supply perspectives.

Energy end-use patterns will converge, even as energy supply structures diverge

The six scenarios indicate that the historical drive toward even more convenient, flexible, and clean fuels reaching the consumer can continue for a wide range of possible future energy supply structures. In all scenarios, there is a shift toward electricity and toward higher-quality fuels, such as natural, oil products, methanol, and, eventually, hydrogen. In contrast to these converging trends, primary energy supply structures diverge, particularly after 2020. Fossil sources continue to provide most of the world's energy well into the next century but to a varying extent across the scenarios. There is a shift away from non-commercial and most unsustainable uses of biomass, and direct uses of coal virtually disappear. Sustainable uses of renewables including modern biomass come to hold a prominent place in all scenarios. They reach the consumer as electricity, liquids, or gases, rather than as solids.

Among regions primary energy supply opportunities vary greatly in the scenarios, with some regional have a much wider range of options than others. This global result was echoed in the regional reviews. NAM has the widest range of possibilities ù wider even that the world as a whole. While recognizing this range of possibilities, the NAM regional review nonetheless expressed clear preferences for "middle-of-the-road" energy development avoiding extreme reliance of particular resources and technologies, whether focused on coal, nuclear, or renewables. The WEU regional review also preferred a diversity of energy-sources, especially to limit import dependence.

At the other end of the spectrum, for MEA all six scenarios are largely dominated by oil and gas, and the regional review identified a clearly preferred scenario : the high-growth, oil and gas intensive Scenario A1. Similarly, PAO, as a high income region with an outstanding record of promoting improved conservation, efficiency, and diversification beyond fossil fuels, is the region best positioned for a high growth transition beyond fossil fuels (i.e. Scenario A3). This global conclusion was largely shared by the regional review. Given PAO's importance as a source of Asian investment, and the importance of early investments, PAO's choices are likely to exert a strong influence on the future of other Asian regions.

Technological change will be critical for future energy systems

Technological change drives productivity growth and economic development, and across all scenarios the role of technological progress is critical. Technological markets grow substantially creating continuing opportunities for improved, adapted, and new technologies. But progress has a price. RD&D of new energy technologies and the accumulation of experience in niche markets required upfront expenditures of money and effort. These are increasingly viewed as too high a price to pay in liberalized markets where the maximization of short-term shareholder value generally takes precedence over longer-term socio-economic development and environmental protection. Yet, it is the RD&D investments of the next few decades that will shape the technology options available after 2020. A robust hedging strategy focuses on generic technologies at the interface between energy supply and end use, including gas turbines, fuel cells, and photovoltaics. These could become as important as today's gasoline engines, electric motors and microchips.

In the regional polls technology was ranked a top priority issue in NAM and PAO, each of which has a particularly strong RD&D and technology base. The PAO regional review emphasized the long-term possibilities of tapping unconventional fossil resources in the form of methane hydrates, an option that will require continued and concerted efforts in technology development. In Sub-Saharan Africa (AFR) the regional review stressed the importance of improved, low-cost technologies tailored to the region's specific circumstances and needs. In CPA the review stressed the importance of technological progress in reconciling rapid economic development and environmental protection.

A key theme of this study is that RD&D and investment choices in the next two decades will lead to technology improvements, and the technologies that benefit will become cheaper, thus making subsequent investments in the same direction more attractive. Given this conclusion, concerns in the WEU regional review that deregulation may curtain incentives for long-term RD&D are particularly important. In both the public and private sectors, recent declines in RD&D budgets need to be carefully checked and a coherent technology strategy developed to respond to long-term challenges. Of all OECD countries, at present only Japan seems prepared to invest in the kind of long-term thinking and technology development that is so vital from the perspective of the global scenario.

Technological change increasingly crosses national borders. One example is that of PAO as the dominant source for both the capital and the technology for most of Asia. However, efforts to transfer technology frequently have failed because successful technology diffusion requires high local expertise. Infrastructures need to be in place, labour must be educated, and local RD&D capabilities are required for the necessary local adaptations, modifications, and extensions of new technologies that make the difference between success and failure. For this reason, capacity building is a resilient strategy for preparing for a transition away from fossil fuels to a more diversified energy supply system. If indigenous technological capabilities are not nurtured right from the outset, however, sunk costs in infrastructure and human capital make it progressively more difficult to change course.

Rates of change in global energy systems will remain slow

The scenarios illustrate that pervasive changes in economic development, energy end-use patterns, and energy supply systems take considerable time. It may take some three decades before FSU's economic restructuring is over and at least five decades before the region catches up to Western Europe's 1990 per capita GDP level. Catching up to WEU standards take AFR and SAS even longer, extending all the way through the 21st century. And even with the ambitious efforts of Case C, it will take almost a century to restructure the global energy system away from its current dominant reliance on fossil fuels.

This was too slow for some reviewers. The CPA and SAS regional reviews, for example, argued for higher rates of both economic growth and energy intensity improvements than those considered feasible in the scenarios. Faster economic growth concurrent with faster energy intensity improvements is consistent with historical experience and the logic underlying the scenarios. The two together do not dramatically alter the evolution of energy demand as described in the scenarios, although faster economic development would mean a developing region could catch up to developed countries more quickly. As emphasized in many regional reviews, however, high growth requires a skilled population, a stable political climate, functioning institutions and markets, free trade, and access to advanced technology.

The CPA regional review also raised the possibility of switching between alternative development paths. For CPA, a preferred alternative scenario would start on a coal-intensive trajectory (Scenario A2) but, after 2020, switch to less carbon-intensive fuels (Scenario A3). Such a transition might be partially possible if carefully planned.

However, from the perspective of the global analysis, changing horses in mid-stream creates two principal difficulties. First, infrastructural and technological investments in coal ultimately will have to be duplicated by new investments in energy alternatives. The railway infrastructure for coal transport, the coal-fired power plants, and the work force dedicated to coal will be in place and expanding, making it ever more difficult to change course. Second, the required massive technological shift would cost more after 2020 than if less carbon-intensive fuels had been the focus of investments right from the beginning. There will not have been the same learning, experience, and cost reductions for the new technologies as would have occurred had they been the original focus of investments.

Interconnectivity will enhance cooperation, systems flexibility and resilience.

Despite energy globalization, market exclusion remains a serious challenge. To date, some two billion people do not have access to modern energy services due to poverty and a lack of energy infrastructures. Many regions are overly dependent on a single, locally available resource, such as traditional fuelwood or coal, and have limited access to the clean flexible energy forms required for economic and social development. Policies to deregulate markets and get "prices right" ignore the poor. Even the best functioning energy markets will not reach those who cannot pay. To include today's poor in energy markets, poverty must be eradicated, and that requires policy action that goes beyond energy policies alone.

What energy policies can accomplish is the improvement of old infrastructures ù the backbone of the energy system ù and the development of new ones. As emphasized in the regional reviews, extended interconnections are needed in Latin America and Africa, and new infrastructures are needed in Eurasia to match the large available resources of oil and gas in the Caspian region and Siberia with the newly emerging centres of energy consumption in Asia. Given FSU's location between the large, mature energy market of WEU and the new, rapidly growing markets of Asia, and given the increasing importance of grid-dependent energy in all scenarios, there are strong incentives for FSU to build a network linking its oil and particularly its gas resources to the large markets to its east and west. This would enhance trade and revenue possibilities enormously while facilitating cleaner, less coal-intensive development in Asia.

Capital requirements will present major challenges for all energy strategies

For all scenarios the capital requirements of the energy sector are large, but not infeasible. Although investment requirements expand more slowly than overall economic growth, the energy sector will have to raise an increasing fraction of its capital from the private sector. It will face stiffer competition and return-on-investment criteria than it has in the past. Moreover, the greatest investment needs are in the new developing world, where current trends in the availability of both international development capital and private investment capital are not auspicious.

Of all the conclusions from the global analysis, the importance of capital requirements was among the most widely endorsed in the regional reviews. In the OECD regions (NAM, WEU and PAO) regional reviewers were most concerned that energy sector deregulation might erode incentives for long-term RD&D and might discourage needed energy infrastructure investments that are big, "lumpy", and risky. In developing regions, regional reviewers emphasized institution reforms to attract the investments needed for expansion. There must be a shift from subsidies and flat rates to segmented market pricing. Losses and leakage must be reduced, and efficiencies improved. Such shifts must be politically acceptable, and they must not leave the poor behind. Market pricing only make an investment attractive if there is a market that can pay the price, and the poor in developing countries do not yet constitute such a market. How quickly they become empowered energy consumers will depend more on the success of policies to reduce poverty than on energy policies.

Regional differences will persist in global energy systems

The future trends central to the scenario results emphasize convergence. Although convergence is sometimes allow relative to current aspirations, in all scenarios economic development is substantial, developing regions narrow the gap between themselves and developed regions, all regions shift toward cleaner, more grid-dependent final energy forms, final energy use per capita converges, energy intensities improve, and decarbonization continues. However, as is abundantly clear from both the regional scenario results and reviews, and the poll results in Table 2, there will continue to be substantial differences among the regions. These arise from differences in populations and natural resource endowments and different economic and technological starting points.

Important regional differences exist, for example, in terms of population, GDP, import dependence, and energy use. Average population growth between 2000 and 2050 varies from a still rapid 2 per cent per year in AFR to even a slight decline in PAO. Average medium-term economic growth (1990 to 2020) in the high-growth Case A ranges between 7 per cent per year in CPA and below 2 per cent per year in both PAO, with its mature economy and aging population, and FSU, whose economy continues to suffer from the recession of the 1990s. GDO per capita in PAO is still between 25 times (Case C) and 35 times (Case B) that in SAS in 2050 (down from nearly 70 times higher in 1990). For CPA, one difference between Scenario A2 and A3 is the difference between being a net energy exporter, in Scenario A2, or net importer, in Scenario A3. MEA and FSU are always net exporters; WEU, PAS and SAS are always net importers. In NAM the post-2020 primary energy structure varies tremendously across scenarios, reflecting the region's diverse resources. In MEA, all scenarios are dominated by oil and gas. Final energy use per capita in 2050 still varies from about 0.6 toe in SAS and AFR to about 5 toe in NAM in Cases A and B (it is closer to 2 toe in Case C in NAM).

However, despite continuing diversity, the consistent message is that the gaps between rich and poor can be narrowed across a wide range of possible energy supply structures. The speed with which they are narrowed will depend on investments in technological progress and infrastructures to integrate regions and reach the poor.

Local environmental impact will take precedence over global change

The natural capacity of the environment to absorb higher levels of pollution, particularly in densely populated metropolitan areas, will become the limiting factor for the unconstrained use of fossil fuels. Local environmental problems are of greater concern to local decision makers than global problems and therefore will have a greater near-term impact on policy. In the developing world indoor air pollution is an urgent environmental problem. A shift away from cooking with wood in open fireplaces will reduce indoor pollution levels currently estimated to be 20 times higher than in industrialized countries. A second urgent problem is the high concentration of particulate matter and sulfur dioxide in many urban areas. Regional air pollution could also prove problematic, especially in the rapidly growing, densely populated coal-intensive economies of Asia. Without abatement measures, sulfur emissions could cause serious public health problems and subject key agricultural crops to acid deposition 10 times sustainable levels.

Agreement that local environmental impacts will take precedence over global change was unanimous in both the regional polls and the detailed regional reviews. It is important, however, not to see the two as separate issues, and to explore strategies that achieve simultaneous progress on both fronts. Improved efficiency and cleaner energy systems (decarbonization) are the key components of the strategy suggested by the global analysis. This emphasis on more efficient and cleaner energy systems as a resilient environmental strategy was invariably echoed in the regional reviews. If policies and investments correspond to the ecologically driven Case C, both regional and local issues are addressed together. But if regional pollution controls are add-ons to fossil-based energy systems such as in Case B, any future reductions in CO_2 emissions only become increasingly remote and difficult. Hence, a number of regional reviews emphasized the need for a long-term perspective and anticipatory RD&D and technology policies to best prepare for future contingencies. After all, climate change is a long-term environmental issue and responses must be prepared in time to prevent disruptions and high costs. This also means that care must be taken not to reduce the flexibility of longer-term energy systems through prematurely locking in existing technologies or exclusively using "add-on" environmental fixes.

Decarbonization will improve the environment at local, regional, and global levels.

Efficiency improvements and structural changes in energy systems, that is, decarbonization, constitute the most resilient strategy in response both to today's local, regional, and global environmental challenges and to those uncertain environmental challenges that may surface in the future. Such a multipurpose strategy was unanimously endorsed in the regional reviews.

The essential message is that quality matters. The trend toward higher-quality, more flexible, and cleaner energy ù an energy mix increasingly composed of electrons and hydrogen rather than carbon atoms ù is strong. As in the past, progress will be gradual. The energy system changes slowly, and big, long-term shifts need to be prepared well in advance. The initiatives, incentives, and policies put into place over the next two decades will determine how quickly these efforts will bear fruit in the form of energy system decarbonization. From the perspective of both the global and regional analysis, a resilient strategy responds to consumer preferences and the need to preserve the environment both locally and globally.

The six scenarios illustrate the congruence of many energy-environmental policies. Efficiency improvements and changes toward cleaner energy conversion and end-use structures not only improve local and regional environments, but also mitigate against undesirable global changes. Consequently, they draw together the urgent local and regional environmental priorities of developing and transitional economies, and the concern about potential climate change evident in many industrialized countries and in countries particularly exposed to potential climate change risks.

But the scenarios also demonstrate that decarbonization is not enough to assure environmental compatibility ù additional active policies will be required. At the regional level, active sulfur abatement strategies will be needed, particularly in the rapidly growing economies of Asia. Were sulfur emissions in the coal-intensive Scenario A2 left unabated, for example, ambient air quality in South and East Asia would deteriorate significantly in both metropolitan and rural areas reaching twice the worst levels ever observed in the most polluted areas of Central and Eastern Europe. For economically important food crops in Asia, Scenario A2's unabated emissions would cause critical loads to be exceeded by factors of up to 10 by 2020.

At the global level, carbon emissions vary substantially among the scenarios, and only the Case C scenarios describe a long-term emission path leading to stabilization of the atmospheric CO_2 concentration below 450 parts per million by volume (ppmv). (Scenario A3 is consistent with stabilization below 550 ppmv of its emissions, which peak around 2060, continue to decline after 2100). In the short-term, only Case C would be unambiguously in compliance with the emissions limited of the December 1997 Kyoto protocol to the UN Framework Convention on Climate Change (FCCC). None of the other scenarios is directly in compliance, but Case B and Scenario A3 come close.

What makes compliance a possibility in Case B and Scenario A3 is that emissions decrease initially in FSU and EEU as a result of economic recessions in the early 1990s. In 2010, FSU emissions are below their Kyoto limits by about 120 megatons of carbon (MtC) in Scenario A3 and 280 MtC in Case B, a difference sometimes referred to as the "Russian bubble". By itself the Russian bubble does not lead to compliance with the Kyoto Protocol. But with international emissions trading the bubble can go a long way in that direction, and Case B and Scenario A3 might therefore also comply with the Kyoto limits. Estimates of the potential trading price of emission reduction units range from US $ 10 per ton of carbon (tC) to US $ 200 per tC, making the total potential value of the Russian bubble huge and raising doubts about the practicality of implementing emission trading. Our analysis suggests that a particularly important potential candidate into which revenues from emission trading might be invested is a new gas transport infrastructure connecting the Siberian and Caspian regions with China and other rapidly developing countries in Asia. This would reduce long-term greenhouse gas emissions and air pollution in Asia and promote a cleaner environment in general.

The Kyoto Protocol is, however, only the first step in a long process. Accomplishing the emissions reductions reflected in Case C would require that timely and tight limits be negotiated efficiently for commitment periods beyond 2008 to 2012, and that limitations are agreed to by all countries.

North and South. Accomplishing stabilization along an alternative path, represented by Scenario A3, might also be possible, for example, if revenues from emissions trading under the Kyoto Protocol are invested prudently in post-fossil technologies and infrastructures. Explicitly dedicating revenues from emissions trading to such environmentally friendly purposes might also make both emissions trading and resource transfers more politically acceptable in the short term.

ANNEXURE - II

17th CONGRESS OF THE WORLD ENERGY COUNCIL—DECLARATION

Introduction

While acknowledging the need for the energy industries to help deal with the effects of current economic and financial turbulence, this Congress focused on the role of technology in meeting the challenges of energy provision and use for the next 50 years.

Delegates to this, the final WEC Congress of the Twentieth Century considered ways in which energy technologies could :

— advance energy supply for those currently without commercial energy services.

— raise efficiency, both in energy provision and use so as to mitigate adverse environmental impacts, including potential climate change, to move towards a sustainable world and to improve the competitiveness of economies.

— reduce local and regional pollution through the wider use of currently available technologies in transport and power generation, and efficient end-use equipment.

— address climate change concerns by deploying more efficient fossil fuel technologies and by using non-fossil fuel energy resources.

— place emphasis on renewable energy resources to supplement conventional resources.

The debate confirmed that technologies cannot of themselves fix all problems. Institutional and policy changes will be required. Finance must be attracted to the right places. Public attitudes and behaviour will need to shift.

Participants outlined current and expected progress in the development of energy technologies and considered the direction energy markets could take in the next 50 years. There was recognition that old certainties and paradigms are giving way to new markets, challenges, organizations and ways of doing things.

As the Exhibition incorporated into the Congress also revealed, the pace of technological development in the energy sector is increasing. Solutions to persistent problems are beginning to emerge with an increased rate of innovative change, encouraged by the current political and business environment of market liberalization, transparent regulation, corporate restructuring and new ways of thinking and working.

The following statement records and issues discussed at the Congress, the Conclusions the WEC has drawn from these discussions and what needs to be done by the WEC and others to increase the availability of sustainable energy for the world's population, and to use this resource with care and efficiency.

The WEC Houston Statement

Key Points

— The world is set for continuing and necessary economic growth, and holds an abundance of accessible energy resources that are more than sufficient to meet this growth.

— A vibrant array of energy technologies are in development that, properly applied, will foster economic development, mitigate environmental impacts, and move in the direction of energy sustainability.

— The liberalization of energy markets, coupled with the right institutional and regulatory framework, is attracting substantial private investment to meet energy needs but the allocation of these funds now and the adequacy of their future flow to the energy sector give grounds for concern.

Nevertheless—

— Current volatility in world currency and equities markets, plus political instability, is having a chilling effect on energy investment in countries where the need is significant. Energy industries need to take a long view and work closely with governments under duress to avoid a downward spiral.

— The problem of world energy poverty persists. Today, as was the case at the last Congress three years ago, one-third of the world's population do not have access to commercial forms of energy, while 20 per cent of the world's population consumes 80 per cent of the world's energy production. Too little progress has been made in addressing those needs. The problems in rural areas are particularly acute and new partnership and economic models are needed to address the problem.

— The possibility of climate change induced by human activity and the ways in which energy production and use can cause severe local environmental degradation and ill health continue to pose enormous challenges for the world's energy sector. As regards climate change, precautionary measures are essential and both developing and developed countries need to shoulder appropriate responsibilities.

Conclusions

The WEC has derived the following conclusions from this Congress :

1. A third of the world's 5.9 billion people do not currently have access to commercial energy. Most of these people live in developing countries where 90 per cent of today's burgeoning population growth is occurring. By 2020, there will be roughly another 2 billion people in the world, mostly in developing countries. The World Energy Council believes that global energy consumption will grow by about 50 per cent in 20 years. Even if the world were organized to use its natural and human resources optimally, this would pose significant challenge.
2. Although current fossil fuel resources are sufficient to sustain global economic growth well into the next century and will be used in increasing amount, severe environmental, economic and technological challenges would follow unless pollution control measures were employed and steps were taken to tackle carbon dioxide and other greenhouse gas emissions. Technologies that reduce net life cycle carbon emissions and sequester carbon must be developed and deployed where economic. Renewable and nuclear energies also should have a role in the necessary decarbonization of the fuel mix.
3. The liberalisation process in the energy sector poses major

challenges for countries with economies in transition and for developing countries. Governments in these countries need to foster transparency, legal certainty and rights of arbitration to underpin the transactional obligations that must be understood by all parties : customers, developers, utilities, stakeholders and regulators.

Sustainable model should be worked out for developing countries looking to the social and financial conditions.

4. While fuelwood and coal will remain the principal energy supply resources for many developing countries, distributed generators ù microturbines, diesel engines, fuel cells, and renewable technologies, specifically wind, biomass and solar technologies, can be expected to provide a viable option for areas operating independent of power grid and fuel pipeline systems. If cost-reduction progress can be sustained, renewables will help in overcoming the problems of energy impoverishment in the developing world.
5. Increased efficiency in the use of energy offers the most immediate, largest and most cost-effective opportunity to reduce resource consumption and environmental degradation, particularly in transportation applications in developed countries. Reducing the energy intensities of economies remains an important goal.
6. Technological innovation is happening quickly and on a wide front to assure clean and efficient use of primary energy resources. Cleaner and more versatile energy forms will characterize future energy systems. Technological breakthroughs are often now achieved by entities outside the existing energy sector with enormous implications for the energy industry. Entering into dialogues with researchers in different disciplines is leading to unexpected synergies, as are collaborative R&D initiatives around the world.
7. As the energy industry is restructured globally, new businesses and alliances are emerging in the upstream production and generation activities as well as in energy distribution. These developments are improving efficiencies in locating and extracting energy resources from deep and remote parts of the globe. New distribution alliances are coupling energy with efficient technologies to provide services customers want, including transportation, lighting, air conditioning and refrigeration services. Information technology is making markets

more efficient, resource production less speculative and costly, and the delivery and monitoring of energy use more effective, while enfranchising customers to make more intelligent choices.

8. Sustainability will depend not only on an adequate energy supply, but also on the stimulation, stewardship and stability that educated and informed populations provide in today's interdependent nations. Populations are no longer isolated by oceans and time zones. The world can view in real time what's making tomorrow's news. Travel, television, the telephone and now the Web serve to integrate interests, promote understanding and create market opportunities. By these means, it is informed customers who are also driving the pace of change.

Recommendations

The WEC, as a result of this Congress, makes the following Recommendations :

1. We urge industry and governments to sustain their efforts to widen the energy supply base into cost effective options; to provide for genuinely clean fossil fuel production and use; to resolve uncertainties related to climate change; to close the nuclear fuel cycle and resolve waste disposal issues, and to pursue economic renewable energy supplies. More study is needed on anthropogenic greenhouse gas emissions and more effort is needed to curb vehicle emissions.
2. While in principle, global energy resources are more than adequate to meet the projected growth in consumption, their location poses political, technological, economic and environmental challenges in both producer and consumer nations. A substantial commitment to the World Trade Organization and market mechanisms to assure optimal access for all countries is essential. Freedom of trade in energy and energy services is an essential feature of ensuring competition and avoiding undue market dominance.
3. The accelerated development and use of appropriate renewable energy resources must be given high priority as a means of supplying commercial energy services to people for the first time. Active solar, biomass, wind and mini-hydro development schemes are endorsed. The WEC should take the lead in modelling new partnerships and new economic systems to

sustain long-term investment programmes designed to bring energy to rural areas.

4. Nuclear power should play a major role in contributing to electricity provision and in strategies to combat global warming. The nuclear industry will have to take the necessary steps to bring down generation costs and to satisfy public concerns about safety. Governments must take a much more active role in assuring prudent regulatory oversight to ensure that nuclear waste is managed properly and the dangers of nuclear proliferation are addressed effectively everywhere.
5. Lifestyles adopted widely in the developed world, where 20 per cent of the world's energy population consume 80 per cent of its energy, appear to run counter to good energy-conservation practice. In this context, WEC has concerns about recent trends that encourage energy-intensive light-duty vehicle use for every day and leisure pursuits. We urge government and industry to assess thoroughly the societal cost of a trend to larger, more energy consumptive options at a time when environmental infrastructure costs are mounting.
6. The WEC recommends auto-oil industry partnerships to reduce local oil pollution, especially in developing economies and to promote the transfer of advanced fuel and automotive technologies to these countries. Policy makers must also focus attention on how best to maintain rail and water-borne freight as an important means of moderating road freight growth, and on future air transportation growth.
7. On the issue of global climate change, the need for immediate cost-effective precautionary measures to mitigate possible climate change is accepted. We remind Parties to the UN Climate Convention of 1992 of the obligations to which they already have agreed. As with the earlier chlorofluorocarbons problem in the refrigeration industry, the "minimum regrets" approach is required now to minimize the collective risk until climate change and anthropogenic greenhouse gas emissions can be linked conclusively. Options to lower greenhouse gas emissions are many, and, for the most part, are not controversial; they should be identified, prioritized and actions taken. Joint implementation and Clean Development Mechanism (CDM) programmes can then be structured to achieve reductions with minimal to tolerate disruptions economically and politically.
8. The energy industry must sustain its commitment to research

and development with government support for activities related to fostering the public good. While proprietary interests will dominate industry efforts, collaborative R&D that extends the breadth, reach and pace of technological progress has important strategic advantages as companies reconfigure their production and customer service businesses. A pronounced decline in government funding of R&D at a time when industry restructuring is lessening private-sector investment in research and development does not bode well for a timely solution to some of the energy industry's long-term objectives. The WEC recommends an increased governmental awareness of the R&D requirements in the energy sector that would substantially enhance the public good.

9. Efficiency should become imperative in every aspect of the energy business from capital allocation and use, through to the location, conversion, transportation and final consumption of energy resources. Government, industry and educational institutions are encouraged to nurture a culture that discourages waste, commits to recycling, and insists on clean, end-use efficient appliances and transportation options.

10. WEC acknowledges that the liberalization of the energy sector globally is an essential step in providing investment for transitioning and developing economies. Caution and care must be given to the regulatory framework and to the ethical and pricing considerations necessary for market-based energy economies. Investors should work with government and international financial institutions to extend the availability of commercial energy to populations in developing nations as rapidly as possible. Energy subsidies should be avoided when possible and minimized where necessary.

11. In industrialized and developing countries, regulatory frameworks should be directed at safeguarding competition and the consumer interest, and not at directing investment or controlling prices. Regulators should also provide a framework that encourages the main actors to take account of the longer-term vision required for sustainable energy provision and use.

12. The WEC will contribute to the implementation of the above Recommendations in the following ways :

— a global electronic energy information system, including regional and national databases, problem solving and technology transfer capabilities in increased collaboration.

— Regional activity and energy facilitation, on a programme rather than a project basis, must increase and can bring real value in terms of energy security, new markets and reductions in emissions. WEC will take the lead in a Programme that commits the world's energy policy makers and industry to provide commercial energy services to 4 billion additional customers by 2020.

— WEC has taken a stand on the business case for the energy world to address greenhouse gas emissions. It will explore the modalities of an industry-supported Pilot Programme for increased collaboration in reducing CO_2 emissions.

WEC will pursue new partnerships. To that end, the WEC has already agreed to conduct a World Energy Assessment jointly with the UN Development Programme (UNDP) and the UN Department of Economic and Social Affairs (UNDESA) as a major contribution to the UN Commission on Sustainable Development and a sustainable energy future for the world. The World Bank also will participate jointly with the WEC in a study on the pricing of energy in developing countries.

ANNEXURE-III

THERMAL GENERATION APRIL-DECEMBER 1998

The total thermal power generation during the period April to December 1998 was 256,781 GWh. SEBs accounted for about 53 per cent of this generation.

SEB	Planned 1998 (GWh)	Actual 1998 (GWh)	1997 (GWh)	% Actual 1998 to Planned 1998	% Actual 1998 to 1997
Andhra Pradesh	14818	14203	13775	96	103
APGCL	468	1322	651	282	203
Assam	988	691	841	70	82
Bihar	1552	1920	1065	124	180
Delhi	2168	1584	1987	73	80
DPL	705	474	701	67	68
Gujarat	20023	16711	16037	83	104
Haryana	2550	2488	2514	98	99
Jammu & Kashmir	44	—	26	0	0
Karnataka	490	416	434	85	96
Kerala	365	188	165	52	114
KPCL	4358	4310	3950	98	109
Madhya Pradesh	13148	12995	12774	99	102
Maharashtra	32460	29943	27830	92	108
Orissa	1920	2044	1881	106	109
Punjab	9234	7544	7703	82	98
Rajasthan	5366	4550	4637	85	98
Tamil Nadu	14674	12718	13178	97	97
Tenughat	1080	1144	993	106	115
Tripura	244	215	130	88	165

Uttar Pradesh	15131	13753	13115	91	105
WBPDCL	4344	5007	4374	115	114
West Bengal	2423	2573	2734	106	94
Total SEB	148580	136793	131495	92	104
Central					
DVC	4717	5398	5020	114	108
NEEPCO	714	642	478	90	134
Neyveli	9523	9698	9760	102	99
NTPC	76274	82269	80748	108	102
Total Central	91228	98007	96006	107	102
Total Private	24002	21981	18132	92	121
All India	263810	256781	245633	97	105

Source: Ministry of Power, Government of India and Powerline.

ANNEXURE - IV

PLANT LOAD FACTOR APRIL-DECEMBER 1998

Thermal Plants registered an average plant load factor of 62.5 per cent during April to December 1998. SEB Plants, on the other hand, reported an average PLF of 58.8 per cent for the same period.

SEB	Planned 1998 (%)	Actual 1998 (%)	1997	% Actual 1998 to Planned 1998	% Actual 1998 to 1997
Andhra Pradesh	73.3	73.3	79.1	100	93
APGCL	—	—	—	—	—
Assam	26.1	18.1	22.2	69	82
Bihar	18.1	22.4	12.4	124	181
Delhi	54.8	37.1	49.7	68	75
DPL	27.4	18.4	27.2	67	68
Gujarat	70.2	62.6	62.3	89	100
Haryana	47.4	46.3	46.7	98	99
Jammu & Kashmir	—	—	—	—	—
Karnataka	—	—	—	—	—
Kerala	—	—	—	—	—
KPCL	79.1	77.7	71.2	98	109
Madhya Pradesh	64.4	63.7	62.6	99	102
Maharashtra	71.3	66.4	66.4	93	100
Orissa	69.3	73.7	67.9	106	109
Punjab	71.5	65.5	68.7	92	95
Rajasthan	80.3	72.7	79.5	91	91
Tamil Nadu	73.8	64.5	66.9	87	96
Tenughat	40.0	41.3	50.5	103	82

Tripura	—	—	—	—	—
Uttar Pradesh	52.9	47.6	46.4	90	103
WBPDCL	52.2	60.2	52.6	115	114
West Bengal	36.2	38.5	40.9	106	94
Total SEB	62.8	58.8	58.9	94	100
Central					
DVC	33.5	37.0	37.6	110	98
NEEPCO	—	—	—	—	—
Neyveli	69.7	71.0	71.4	102	99
NTPC	68.0	72.1	72.2	106	100
Total Central	64.5	68.0	68.5	105	99
Total Private	67.8	70.3	73.7	104	95
All India	63.6	62.5	62.9	98	99

Source: Ministry of Power, Government of India and Powerline.

ANNEXURE - V

IPP PROJECTS

With capacity of 250 MW and above

There are 78 IPP projects each with a capacity of 250 MW and above which are still alive on paper. These add up to a capacity of over 49,600 MW, accounting for almost 80 per cent of the proposed IPP capacity alive on paper in the country. These projects are spread across the state. However, Tamil Nadu and Madhya Pradesh account for a big chunk.

Location	State	Capacity	Fuel	Promoter
Visakhapatnam	Andhra Pradesh	1040	Coal	Hinduja-National Power Corp.
Krishnapatnam B	Andhra Pradesh	520	Coal	BB! Power Inc.
Ramagundam	Andhra Pradesh	520	Coal	BPL Power Projects
Krishnapatnam A	Andhra Pradesh	520	Coal	GVK Power
Vemagiri	Andhra Pradesh	468	Gas/Naphtha	Ispat Power
Kondapalli	Andhra Pradesh	355	Gas/Naphtha	Kondapalli Power Corp.
Peddapuram	Andhra Pradesh	300	Gas/Naphtha	Gautami Power
Bawana	Delhi	421	Gas/Naphtha	Reliance Power
Narela	Delhi	300	Coal	Apollo Energy
Paguthan	Gujarat	655	Gas/Naphtha	Gujarat Torrent Energy Group
Hazira I	Gujarat	515	Gas/Naphtha	Essar Power
Hazira II	Gujarat	515	Gas/Naphtha	Essar Power
Savli	Gujarat	500	Refinery residue	Indian Oil Corp
Vadinar	Gujarat	500	Petroleum coke	Vadinar Power
Panipat	Haryana	300	Refinery residue	Indian Oil Corp
Kinnaur	Himachal Pradesh	1000	Hydel	Jaiprakash Hydro Power

Kinnaur	Himachal Pradesh	300	Hydel	Jaiprakash Hydro Power
Almatti	Karnataka	1100	Hydel	Chamundi Power Corp.
Mangalore	Karnataka	1000	Coal	Mangalore Power Co.
Chamalapura	Karnataka	1000	Coal	Mysore Power Gen.
Nandikur	Karnataka	1000	Coal	Nagarjuna Power Corp.
Vajarahalli	Karnataka	500	Gas/Naphtha	Pulakeshi Power Co.
Raichur	Karnataka	500	Coal	Kutub Power Co.
Bijapur	Karnataka	420	Coal	KEI Energy
Dharwar	Karnataka	350	LNG	Innox Power
Bodadi	Karnataka	300	Gas/Naphtha	KPC Bidadi Power Corp.
Vypeen	Kerala	300	Gas/Naphtha	Siasin Energy
Cheemeni	Kerala	1200	Gas/Naphtha	BPL Power
Kannur	Kerala	516	Gas/Naphtha	Kannur Power Projects
Ambalamugal	Kerala	513	Refinery residue	Cochin Refineries
Kasargod	Kerala	500	Gas/Naphtha	Finolex Energy Corp.
Kottukal	Kerala	495	Gas/Naphtha	Kumar's Energy Corp.
Palakkad	Kerala	348	Gas/Naphtha	Palakkad Power Generating Co.
Korba East	Madhya Pradesh	330	Coal	Daewoo Power
Bina	Madhya Pradesh	1070	Coal	Bina Power Supply Co.
Bhilai	Madhya Pradesh	578	Coal	Bhilai Power Supply Co.
Raigarh	Madhya Pradesh	574	Coal	Jindal-Ganting Power
Korba West	Madhya Pradesh	550	Coal	RPG Group
Pench	Madhya Pradesh	520	Coal	Indian Oil Corp.
Bilaspur	Madhya Pradesh	500	Coal	Indian Thermal Power

Mandaleshwar	Madhya Pradesh	500	Hydel	Shree Maheshwar Hydel Power
Jhabua	Madhya Pradesh	420	Gas/Naphtha	Kedia Power
Guna	Madhya Pradesh	400	Gas/Naphtha	STI Power India
Raigarh	Madhya Pradesh	360	Gas/Naphtha	Alpine Power Systems
Bhander	Madhya Pradesh	347	Gas/Naphtha	Essar Power
Bilaspur	Madhya Pradesh	343	Coal	Madhya Bharat Power Corp.
Ratnagiri	Maharashtra	330	Gas/Naphtha	Dabhol Power Co.
Bhadrawati	Maharashtra	250	Coal	Central India Power Co.
Patalganga	Maharashtra	2450	Gas/Naphtha	Reliance Power
Khaperkheda	Maharashtra	1082	Coal	Ballarpur Industries
Hirma	Orissa	447	Coal	Consolidated Electric Power Asia
Jharsuguda	Orissa	420	Coal	AES Transpower
Durgapur	Orissa	3960	Coal	JK Corp.
Duburi	Orissa	500	Coal	Kalinga Power
Gowindwal Sahib	Punjab	500	Coal	GVK Industries
Bhatinda	Punjab	500	Coal	PSEG
Dholpur	Rajasthan	500	Gas/Naphtha	RPG Dholpur Power Co.
Kapurdi	Rajasthan	703	Lignite	West Power
Jalipa	Rajasthan	600	Lignite	West Power
Barsingsar	Rajasthan	600	Lignite	Hindustan Vidyut Corp.
Barmer	Rajasthan	500	Lignite	DLF Power
Vembar	Tamil Nadu	250	Gas/Naphtha	Indian Power Projects
Cuddalore	Tamil Nadu	2000	Coal	Cuddalore Power Co.
Ennore	Tamil Nadu	1320	Coal	Videocon Power
Kattupalli	Tamil Nadu	1050	Gas/Naphtha	Chennai Power Generation

Tuticorin	Tamil Nadu	1000	Coal	Spic Electric Power Corp.
Ennore	Tamil Nadu	525	Coal	Tri-Sakthi Energy
Jayakondam	Tamil Nadu	525	Lignite	Reliance Power
Pillaiperumanallur	Tamil Nadu	500	Gas/Naphtha	PPN Power Generating Co.
Srimushnam	Tamil Nadu	330	Lignite	TICAPCO
Neyveli	Tamil Nadu	250	Lignite	ST-CMS Electric Co.
Etah	Uttar Pradesh	250	Coal	Jawaharpur Power
Shahjahanpur	Uttar Pradesh	800	Hydel	Rosa Power Supply Co.
Vishnuprayag	Uttar Pradesh	567	Gas/Naphtha	Jaiprakash Hydro Power
Kosi Kalan	Uttar Pradesh	400	Hydel	Wasan Power
Pauri Garhwal	Uttar Pradesh	355	Refinery residue	Duncans North Hydro
Haldia	West Bengal	330	Coal	Indian Oil Corp.
Mouza Muthaberia	West Bengal	700		Bakreshwar Power
		420		

Source: Ministry of Power, Government of India and Powerline.

ANNEXURE - VI

SOURCE OF POWER (BY OWNERSHIP)

Eight states (AP, Gujarat, Karnataka, MO, Maharashtra, Punjab,TN and UP) account for over three-fourths of the total installed capacity in the country, MSEB's installed capacity is the highest among the SEBs at 8,231 MW. Meanwhile, states are also entitled to power from central sector units. Moreover, in some states like Gujarat and AP, IPP power is also available.

SEB	Total Board	Share from Gentral Units	Inter state Share	Others*	Total****
Andhra Pradesh	5885	897	0	495	7277
Assam	574	294	0	25	893
Bihar**	1989	928	0	0	2917
Delhi	678	1679	0	720	3077
Gujarat	4690	1293	0	1529	7512
Haryana	863	613	917	0	2393
Himachal Pradesh	300	247	133	61	741
Jammu & Kashmir	372	0	0	0	372
Karnataka***	3579	700	0	3	4282
Kerala	1785	574	0	12	2371
Madhya Pradesh	3814	1797	0	0	5611
Maharashtra	8231	1543	190	2274	12238
Meghalaya	189	150	0	0	339
Orissa	1692	1147	0	150	2989
Punjab	2245	1080	1281	10	4616
Rajasthan	1052	1104	941	694	3791
Tamil Nadu	5075	1841	0	0	6916
Uttar Pradesh	6049	2714	0	0	8763
West Bengal***	2891	820	0	0	3711

* Includes IPP capacity. ** Includes Tenughat Vidyut Nigam Ltd.

*** Includes Karnataka Power Corporation. ****Includes WBPDCI, DPL and DPSC.

SOURCE OF POWER (GENERATION/PURCHASE)

SEBs accounted for about 55 per cent of the total generation in the country in 1997-98. They also purchase power from other SEBs, IPPs, and central and private utilities. Some states, like Delhi and Bihar, are more reliant on purchase from other sources. Other states, like Maharashtra and Andhra Pradesh, are more reliant on their own generation.

SEB	1996-97(%)		1997-98 (%)	
	Generation	Purchase*	Generation	Purchase
Andhra Pradesh	75	25	69	31
Assam	49	51	38	62
Bihar**	20	80	NA	NA
Delhi	16	84	NA	NA
Gujarat	67	33	65	35
Haryana	28	72	29	71
Himachal Pradesh	41	59	39	61
Jammu & Kashmir	20	80	NA	NA
Karnataka***	65	35	76	24
Kerala	63	37	55	45
Madhya Pradesh	59	41	61	39
Maharashtra	76	24	74	26
Meghalaya	86	14	88	12
Orissa	36	64	36	64
Punjab	77	23	72	28
Rajasthan	52	48	49	51
Tamil Nadu	70	30	68	32
Uttar Pradesh	63	37	63	37
West Bengal***	30	70	32	68

* Percentatge of total power available (generation plus purchase).

** Includes Tenughat Vidyut Nigam Ltd.

*** Includes Karnataka Power Corporation.

**** Includes WBPDCL, DPL and DPSC.

INDEX